ATMOSPHERIC DISPERSION MODELLING

ATMOSPHERIC DISPERSION MODELLING

An Introduction to Practical Applications

Rod Barratt

Routledge
Taylor & Francis Group
LONDON AND NEW YORK

First published 2001 by Earthscan

2 Park Square, Milton Park, Abingdon, Oxon OX14 4RN
711 Third Avenue, New York, NY 10017, USA

Routledge is an imprint of the Taylor & Francis Group, an informa business

First issued in hardback 2016

ISBN 13: 978-1-85383-642-8 (pbk)
ISBN 13: 978-1-138-15724-8 (hbk)

Cover design by Andrew Corbett

A catalogue record for this book is available from the British Library

Library of Congress Cataloging-in-Publication Data

Barratt, Rod, 1947-.
Atmospheric dispersion modelling : an introduction to practical applications / Rod Barratt.
p. cm. – (Business and environment practitioner series)
Includes bibliographical references and index.
ISBN 1-85383-642-7 (pbk.)
1. Air Pollution-Mathematical models. 2. Atmospheric diffusion-Mathematical models.
[DNLM: 1. Air Pollutants, Environmental-analysis. 2. Air Pollutants, Environmental-standards. 3. Environmental Monitoring-methods. 4. Models, Theoretical. WA 754 B271a 2001] I. Title. II. Series

TD883 .B267 2001
628.5'3'015118-dc21

2001005490

CONTENTS

LIST OF ILLUSTRATIONS

Figures

Tables

ACKNOWLEDGEMENTS

The idea for this book was prompted by the aversion that many students show to atmospheric dispersion modelling. This suggested a need for a book that introduced some of the practical aspects of this increasingly important technique, but without the detailed mathematical treatment covered in many other texts on the subject.

I am grateful to Earthscan for giving me the opportunity to write such a book, and to Dr Ruth Hillary who provided valuable guidance when I embarked on the task and advice for improving the structure. Feedback from other critical readers also helped to improve the early drafts, while Akan Leander managed the production process.

I am indebted to several other people who have helped with various aspects of this book. Jere Folgert, Linda Johnson and Bob Appleby kindly allowed me to use images from their own modelling activities, while my daughter Elisabeth prepared several of the diagrams. Gratitude is also expressed to companies that have permitted me to show images from their software. However, this does not constitute the endorsement of specific products, but is rather an illustration of the range that is available to the modelling community. Models are continually being improved and enhanced, so specific versions are not mentioned here.

Where I have drawn on published material, I have provided appropriate references in footnotes for readers who wish to delve further into this extensive and developing area. Some additional sources of information I have found useful, including internet addresses, are included in the appendices.

Finally, I must acknowledge the patience and tolerance of my family during the time it took to complete this book.

SUMMARY

'... we may be allowed to repute some Aers pure, comparatively, viz. That which is cleare, open, sweetly ventilated and put into motion with gentle gales and breezes...'*

Air quality is far from being a new concern but remains an issue demanding increasing attention. The dispersion and ground-level concentration of pollutants are determined by the complex interactions of many factors. These include meteorological conditions at or near the site, the topography of the surrounding areas, physical characteristics of stack or other emission points as well as the physical and chemical characteristics of the pollutants. Many measurements of the concentrations of specific air pollutants at selected monitoring locations are made throughout the world. However, there is always a need to predict future air pollution levels or to estimate air quality where no monitoring data are available. Computer models are a relatively inexpensive and expedient way of providing such air quality information. They also provide an important planning tool because of their ability to evaluate a variety of options for managing air quality. Projects that either introduce new sources of emissions or aim to reduce emissions require careful scrutiny to assess any risks as far as possible. By employing tools such as modelling and exposure assessment, risk based decision-making considers the current and potential risks posed by a release. This knowledge is then used to make decisions about corrective action required for processes, as well as implications for site management and emergency response. For many sources, this will typically require mathematical modelling of the changes in ambient concentrations that result from the changes.

Official guidelines from government bodies note that the selection of models and the interpretation of their results are not straightforward. This book aims to clarify some of these points. Practical examples are included throughout, and illustrations of modelling software are integrated in the text. Further aims of this book are as follows:

- Describe the characteristics influencing the transport of harmful materials through the air.
- Outline the general theoretical principles underpinning dispersion modelling.
- Introduce several types of model used for air quality management.
- Identify the limitations of dispersion models.
- Provide the skills to use simple dispersion models.
- Enable the reader to understand and interpret the output from models of real problems involving continuous or accidental releases to the atmosphere.

The first four chapters place modelling in the context of the current practice of air quality management, and outline sufficient theoretical background to allow the novice modeller to understand the requirements and limitations of input data. Chapter 5 outlines basic features of the Gaussian plume model, although other books go into more theoretical detail on this. The next three chapters follow the staged approach to modelling, starting with simple models and progressing to refined models for both continuous and emergency releases. Chapter 9 focuses on how models match reality and illustrates validation issues, while guidelines for good practice conclude the book.

As the title implies, this book is intended to offer a practical introduction to atmospheric dispersion modelling. The serious mathematician is encouraged to explore other books that deal with the theoretical background, while practitioners will need to focus on issues relevant to their own models and applications.

* John Evelyn, *Fumifugium: Or the Inconvenience of the Aer and Smoake of London Dissipated*, 1661, reprinted by the National Society for Clean Air, 1961

ABOUT THE AUTHOR

A Chartered Chemist by profession, Dr Rod Barratt has spent most of his career in the teaching and practice of air quality management. He built a practical foundation in his subject in local government environmental protection and engineering consultancy. In 1987 he joined the Open University, where he is Head of the Department of Environmental and Mechanical Engineering. His teaching interests focus on air quality management and wider aspects of safety, health and environmental management.

In addition to about 40 journal publications, he has written two books dealing with environmental management, and has contributed a chapter on energy management to a book on clean technology. As an expert witness, he has used atmospheric dispersion modelling in developing evidence for public inquiries relating to the planning aspects of road, industrial and mineral working activities.

ACRONYMS AND ABBREVIATIONS

ADMS	Atmospheric dispersion modeling system
ADPIC	Atmospheric Diffusion Particle-in-cell Model
AERMIC	American Meteorological Society/EPA Regulatory Model Improvement Committee (US)
AERMOD	AERMIC dispersion Model
AQMA	Air quality management area
AQS	Air quality standards
ARCHIE	Automated Resource for Chemical Hazard Incident Evaluation
ARN	Automatic rural network
ATMES	Atmospheric Transport Model Evaluation Study
AUN	Automatic urban network
BAT	Best available techniques
BATNEEC	Best available techniques not entailing excessive costs
BLEVE	Boiling liquid expanding vapour explosion
BPEO	Best practicable environmental option
BUN	Basic urban network
CAAA	Clean Air Act Amendments, 1990 (US)
CAR	Calculation of air pollution from road traffic model
CFD	Computational fluid dynamics
CHARM	Complex Hazardous Air Release Model
COMEAP	Committee on Medical Aspects of Air Pollution
DEGADIS	Dense gas dispersion model
DEM	Digital Elevation Model
DMRB	Design Manual for Roads & Bridges
DNMI	Norwegian Meteorological Institute
DREAM	Danish Rimpuff and Eulerian Accidental Release Model
EAL	Environmental assessment level
EDMS	Emissions and dispersion modelling
EEA	European Environment Agency
EMAS	Eco-management and audit scheme
EPA	Environmental Protection Agency (US)
EPAQS	Expert Panel on Air Quality Standards
EPRI	Electric Power Research Institute (US)
EXTEX	Long Range Dispersion Model Evaluation
FAA	US Federal Aviation Administration
GUI	Graphical user interface
HMIP	Her Majesty's Inspectorate of Pollution (now part of the Environment Agency)
HSE	Health and Safety Executive (UK)
HYsplit	Hybrid Single Particle Lagrangian Integrated Trajectory
IAEA	International Atomic Energy Agency
IPC	Integrated pollution control
IPPC	Integrated pollution prevention and control
ISC	Industrial Source Complex model
LADM	Lagrangian Atmospheric Dispersion Model
LAQM	Local air quality management
MACT	Maximum achievable control technology (US)
MATHEW	Mass-adjusted Three-dimensional Model

MEL	Maximum exposure limit
NAEI	National Atmospheric Emission Inventory
NAME	Nuclear Accident Model
NCDC	National Climatic Data Center (US)
NERI	Danish National Environmental Research Institute
NOA	National Observatory at Athens
NOAA	National Oceanic and Atmospheric Administration (US Department of Commerce)
NTIS	National Technical Information Service (US)
OEL	Occupational exposure limit
OES	Occupational exposure standard
OML	Operationalle Meteorologiske Luftkvalitetmodeller (Denmark)
PAH	Polycyclic aromatic hydrocarbon
PCB	Polychlorinated biphenyl
PHAST	Process Hazard Analysis Software Tools
PMCH	Pefluoro-methyl-cyclohexane
ppb	Parts per billion (10^9)
ppm	Parts per million
RAINS	Regional air pollution information and simulation model
RIVM	Netherlands National Institute of Public Health and Environmental Protection
SMHI	Swedish Meteorological and Hydrological Institute
STEL	Short term exposure limit
TOMPS	Toxic organic micropollutants
TOXST	Toxic Modelling System Short Term model
TSCREEN	Toxic Screening model
USGS	US Geological Survey
WHO	World Health Organization
WMO	World Meteorological Organization

CHAPTER 1
WHY MODEL AIR QUALITY?

The aims of this chapter are to introduce air quality modelling and to justify its role as a tool for environmental management in the context of legal and commercial pressures.

There is a popular misconception that air pollution simply disperses over time into the vastness of the atmosphere. Sadly, this is a fallacy, and effects at a local, regional, continental and even global scale are possible. Numerous incidents have occurred and continue to occur to demonstrate the truth of this statement. Understanding the behaviour of substances released into the atmosphere can help in designing processes with pollution prevention in mind. If the worst happens and an incident occurs, knowledge of atmospheric transport may help to mitigate the effects.

Mathematical models can represent the behaviour of a release into the atmosphere. Modelling offers a means to explore the impacts in a manageable way. The aim of this book is not to cover the pure science and mathematics of modelling, which are well covered elsewhere with some sources identified in the bibliography. Instead, it aims to introduce the reader to modelling, and to develop the capability to use or understand the results of models for air quality management and emergency response planning.

The origins of modelling lie in the need to predict the transport and dispersion of radionuclides generated by nuclear research in the 1950s. Since that time, the relevance of modelling has extended to other areas, ranging from photochemical smog formation and acid deposition to emergency response. Other applications of models range from resolving problems associated with pesticides released by surface or aerial spraying, odours associated with the land application of animal waste, as well as releases from the open burning of energetic and other materials such as animal carcasses during the foot and mouth outbreak in 2001 in the UK. However, a dominant feature in many countries has been the need for the prediction of atmospheric transport and dispersion for regulatory purposes, especially in terms of permits, licences or authorizations to operate. Models are invariably used by expert witnesses in planning inquiries where air pollution is an issue.

Why Model?

In any urban area there may be many different sources influencing air quality (Figure 1.1). These sources include traffic and large industrial sources as well as smaller emissions from local businesses and domestic installations. In some places, local sources may be important, for example, in a street canyon. Elsewhere, the sources of pollutants making the greatest contribution to pollutant concentrations may be distant. Under the Environment Act 1995, in the UK local authorities have a statutory duty to review and assess air quality. In the terminology of the UK Air Quality Limit Values Regulations 2001* 'assessment' means any method used to measure, calculate, predict or estimate the level of a relevant pollutant in the ambient air. Likewise, in the US source permitting demands the prediction of environmental impact. Computer based dispersion modelling is one of the tools available to assist in these processes.

---

* The Air Quality Limit Values Regulations 2001, Statutory Instrument 2001 No. 2315, Her Majesty's Stationery Office, London.

Figure 1.1 Air Quality Management

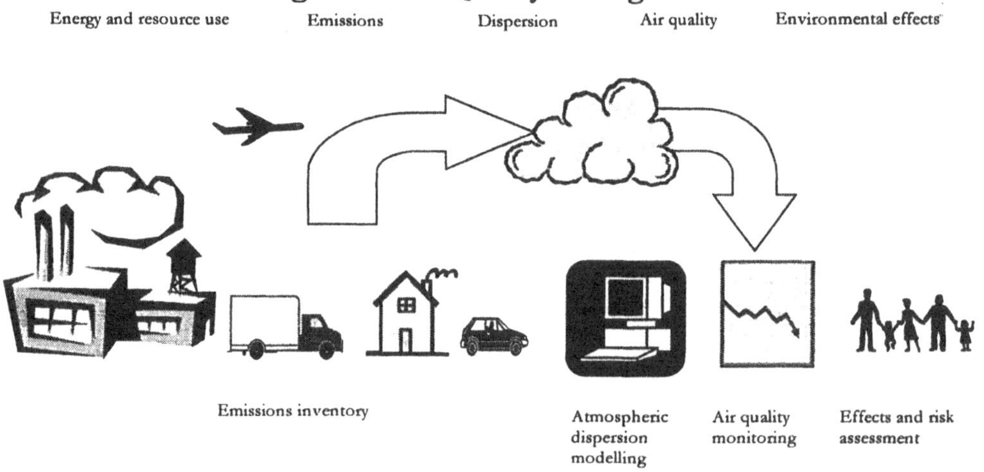

Modelling complements air quality monitoring by providing information that cannot be provided by other means. For example, it allows:

- estimation of air quality across a wide area rather than at a few locations;
- predictions of the contribution by specific sources to pollution at locations of interest; and
- predictions of future air quality, especially resulting from sources that currently do not exist.

A Complement to Monitoring

Rational decision-making for environmental management requires a firm basis in data on air quality. Often, it may be appropriate to acquire information on nearby air quality, perhaps to relate to operations at a site or to compare the predicted impact of future developments against current air quality. One may even wish to compare indoor air quality with that outdoors. The growing number of continuous air quality measuring stations in the UK means that an overwhelming mass of aerometric data are readily available from the networks and archives. Similar data are produced in other countries, and some routes to these 'online' are given in Appendix 5.

Environmental quality data may be useful when an organization needs information on local environmental quality, perhaps when assessing the impact of its existing or proposed activities on the wider environment. However, due to limitations in the spatial and temporal coverage of air quality measurements, monitoring data normally are not sufficient as the single means of demonstrating the adequacy of emission limits for existing sources. In addition, the impacts of new sources that do not yet exist can only be determined through modelling. Therefore, while uniquely filling one need, models have also become a primary analytical tool in many air quality assessments. Air quality measurements complement dispersion models, with each technique having strengths and weaknesses. Measurements are particularly useful in assessing the accuracy of model estimates as will be described later in the context of validation. The use of air quality measurements alone, however, could be preferable when models are found unacceptable and monitoring data with sufficient spatial and temporal coverage are available. As will be emphasized later, it is essential that predicted additional pollution from a source is added to background levels in an area to estimate overall air quality.

Satisfying Legal Pressures

Modelling of the dispersion of components released into the atmosphere provides information for the decision-making process in several areas of risk assessment, including plant design, setting operating conditions, comparing options for site location and the estimation of exposures resulting from releases. In addition to the applications that may serve organizational needs, there are also regulatory requirements for which modelling may be useful.

The framework for improving air quality in the UK is currently set by Part IV of the Environment Act 1995. Two main elements of the Act are:

- A requirement to produce a National Air Quality Strategy containing standards, objectives and measures to achieve the objectives.
- A local air quality management system (LAQM) giving statutory duties for local authorities to review their air quality, assess it against air quality objectives set in regulations and, where it is likely the objectives will not be met, declare air quality management areas and draw up and implement appropriate action plans.

The strategy, first published in March 1997 and most recently revised in 2000, contained standards and policy objectives for eight pollutants, based on advice from the Expert Panel on Air Quality Standards (EPAQS). The Air Quality Limit Values Regulations 2001 (Appendix 1) updated the legal status of air quality objectives that enable local authorities to undertake their statutory duties. The EU Air Quality Framework and Daughter Directives place legal obligations on member states to achieve the limit values for individual pollutants by specified dates. Hence the limit values are incorporated into the national strategy, and are specified for the purposes of LAQM, so that the necessary local action is taken to meet the obligations. The appendices also include US standards for comparison as well as World Health Organization standards and guidelines.

A system of air quality banding used in the UK since 1990 was revised in 1998 to take into account the then air quality standards. The revised system, like the old, is intended to provide guidance on the effects of air pollutants on health. The bands are briefly indicated in Table 1.1.

Table 1.1 Air Quality Banding in the UK

Level of air pollution	Ozone Running 8 hr/1 hr average		Sulphur dioxide 15 min average		Nitrogen dioxide 1 hr average		Particulate matter* Running 24 hr average	Carbon monoxide Running 8 hr average	
	ppb	$\mu g.m^{-3}$	ppb	$\mu g.m^{-3}$	ppb	$\mu g.m^{-3}$	$\mu g.m^{-3}$	ppm	$mg.m^{-3}$
'low air pollution'	< 50	< 100	< 100	< 266	< 150	< 287	< 50	< 10	< 11.6
'moderate'	50 – 89	100 – 179	100 – 199	266 – 531	150 – 299	287 – 572	50 – 74	10 – 14.9	11.6 – 17.3
'high'	90 – 179	180 – 359	200 – 399	532 – 1063	300 – 399	573 – 763	75 – 99	15 – 19.9	17.4 – 23.1
'very high'	≥ 180	≥ 360	≥ 400	≥ 1064	≥ 400	≥ 764	≥ 100 $\mu g\ m^{-3}$	≥ 20	≥ 23.2

*Because of the continuous relationship between concentrations of particles and effects on health a different approach to devising bands of air quality was advised by the Committee on the Medical Aspects of Air Pollution (COMEAP).

There are proposals to subdivide the four descriptor bands into a numerical index in which low is the range 1 –3, moderate covers 4–6, high is 7–9 and 10 indicates very high levels of air pollution.

Objectives for gaseous pollutants may be expressed in terms of parts per billion (ppb) or parts per million (ppm), and particulate pollutants (PM_{10} and lead) in terms of mass per unit volume (μg m^{-3}). In order to avoid confusion between different approaches from different bodies, the UK government decided to express all objectives in its strategy in terms of mass per unit volume with, where appropriate, the equivalent ppb or ppm figure in brackets. Thus, conversions between ppb or ppm and μg m^{-3} and mg m^{-3} are at 20°C and 1013 mb, unless otherwise stated. Conversions are outlined in Appendix 1.

Environmental Assessment Levels

For many substances released to the ambient atmosphere there are no environmental quality standards. For the purposes of the former regime of integrated pollution control (IPC) under the provisions of the Environmental Protection Act 1990, the concept of a regulatory limit known as an Environmental Assessment Level (EAL) was introduced. An EAL is the concentration of a substance in a particular environmental medium that the regulator uses for comparison purposes to assess significant releases. In the absence of a comprehensive list of EALs, interim values were based on a variety of data sources including EPAQS, European Union (EU) Guidelines, WHO Guidelines, other international guidelines (for example from the UN Economic Commission for Europe), Health and Safety Executive (HSE) occupational exposure standards (OES) or maximum exposure limits (MEL) and expert judgement.

The Environment Agency regulates over 400 substances or groups of substances with many being potentially released to the atmosphere. The sets of criteria above deal with only a few of these substances. The practice of using OEL values has a long history of application to calculations of critical levels of pollutants in the determination of appropriate stack heights.

The use of occupational standards was addressed in UK regulatory guidance.* Table 1.2 gives examples for releases to air.

Table 1.2 Environmental Assessment Levels

	Long term EAL		Short term EAL	
	μg.m^{-3}	source	μg.m^{-3}	source
ammonia	100	EPA (a)	2400	EH40 (OES)
arsine	2	EH40 (OES)	60	EH40 (OES)
benzene	3.25	EPAQS	960	EH40 (MEL)
Cd and compounds	0.005	WHO	1.5	EH40 (MEL)
carbon monoxide	550	EH40 (OES)	99	EH40 (OES)
chlorine	1.5	EH40 (OES)	300	EH40 (OES)
ethanol	19000	EH40 (OES)	570000	EH40 (OES)
ethylene oxide	20	EH40 (MEL)	600	EH40 (MEL)
formaldehyde	5	EH40 (MEL)	100 (b)	WHO
hydrogen sulphide	140	EH40 (OES)	150 (c)	WHO
n-octane	14500	EH40 (OES)	0.02	EH40 (OES)
styrene	840	EH40 (MEL)	800 (c)	WHO
tetrachloroethylene	3350	EH40 (OES)	5000 (c)	WHO
trichloroethylene	1070	EH40 (MEL)	1000 (c)	WHO

(a) US Environmental Protection Agency
(b) denotes 24 h reference period
(c) denotes 30 min reference period

Derivation of these guidelines is illustrated as follows. The exposure of members of the local population to contaminants of their immediate atmospheric environment is assumed to be a

* HMIP Technical Guidance Note (Environment) E1, *Environmental, Economic & BPEO Assessments*, Volume III, Technical Data, 1995

maximum of 7 x 24 hrs = 168 hrs each week and this may occur over 52 weeks of a year. By comparison, the working exposure is judged to be 5 x 8 hrs = 40 hrs each week over 44 weeks. This comparison suggests that a safety factor for converting an OEL to an EAL should be:

$$(168/40) \times (52/44) = 4.96.$$

However, the essentially continuous exposure of the general population does not allow for a recovery period over an exposure period that may be a lifetime. A further safety factor of two gives a total factor of ten to divide an occupational standard to generate a value for an EAL.

In addition, there is the likelihood that the working population is the relatively 'healthy' group, whereas the general population includes children, the elderly and the sick, including those with illnesses such as asthma and who are sensitive to airborne contaminants. By convention a further safety factor of 10 allows for the differences between these groups of people, giving a combined factor of 1/100 of the 8 hr occupational limit. Just as there is a distinction between the two types of long term occupational exposure limits as occupational exposure standards (OES) and maximum exposure limits (MEL), so a distinction was made in the derived EAL for any substance for which a MEL applies. An additional safety factor of five was applied.

While these derived limits for environmental assessment levels have been subject to public scrutiny, they are not based on robust toxicological data. So, they must still be treated with caution and are subject to review.

Wider Pressures

Some obligations come from EU imperatives on industrial pollution control and risk management. The relevant EU directives and regulations cover four areas, namely, environmental impact assessments, the control of industrial emissions, the control of major accident hazards and pressures for environmental audits. The first area includes assessing environmental impact for land use planning (for example as required under EC Directive 85/337/EEC and resulting in national legislation such as the Town and Country Planning (Environmental Impact Assessment) Regulations 1999. The second area includes the Integrated Pollution Prevention and Control Directive 96/61/EEC (IPPC) that established requirements for permits for the operation of certain industrial activities to control releases to air and water, and wastes. This is implemented in the UK through the Pollution Prevention and Control Act 1999 and associated regulations. Similar principles are found elsewhere. For example, in the US, Title III of the Clean Air Act Amendments 1990 (CAAA) sets out a framework for regulating major sources of hazardous air pollutants. Under this framework, the US Environmental Protection Agency (EPA) is obliged to assess the residual health risks to the population near each regulated source. The measurement of long term and short term ambient concentrations for each toxic air pollutant (189 pollutants as listed in Section112(b)) near each source would be prohibitively expensive. However, modelling offers a means towards determining the residual risk by predicting concentrations.

The third area, the control of major accident hazards, covers the Seveso II Directive (96/82/EC) named after the town in Italy that in 1976 was the site of a major accidental release of toxic gas. This directive replaced an earlier one, which was a model for similar legislation outside Europe, and required operators of industrial plants to identify major accident hazards and to take steps to control them and to limit their effects. The Bhopal tragedy in India inspired amendments to the Seveso I Directive as well as the US Clean Air Act 1990 requirements that businesses develop plans to prevent accidental releases of toxic chemicals. The Seveso II Directive strengthened the requirements for safety management and emergency planning by operators of certain industrial plants. Operators of the industrial installations concerned must prepare a safety report and an internal emergency plan. Modelling has a role here in a similar manner to the requirements under US legislation. Off-site consequence analysis is part of risk management planning in which modelling has a role. Other incidents, such as that at Chernobyl further support a need for modelling to aid decision-making on dealing with the impact of disasters.

Environmental Management Systems

Beyond legal pressures there are voluntary schemes for improving environmental performance through the introduction of environmental management systems. The British Standard, BS 7750 was the first such scheme, but was withdrawn in April 1997 following the introduction of the International Standard, ISO 14001. Regulation 1836/93/EEC set up a voluntary eco-management and audit scheme (EMAS) for participating industrial companies that seeks to reward and promote better environmental performance. Subsequent revision of EMAS by Regulation (EC) 761/2001 embedded ISO 14001 in its structure.

All of the options available to organizations to introduce formal environmental management systems include a need to consider their environmental effects. Some issues to be considered are as follows:

- How *significant* or severe are the potential environmental impacts should a process failure occur?
- How *frequently* will the situation arise that could lead to the impact?
- What are the *significant* environmental aspects, considering impacts, likelihood, severity and frequency?
- Are the *significant* environmental impacts local, regional or global in scope?

Reflecting back on Figure 1.1 it is evident that the relationship between environmental aspects and impacts is one of cause and effect. An environmental aspect refers to an element of an organization's activity, product or service that can have a beneficial or adverse impact on the environment. For example, it could involve a discharge, an emission, consumption or reuse of a material or noise. An impact refers to the change that takes place in the environment because of the aspect. Examples of impacts might include odour or other air pollutants causing a nuisance to neighbouring premises or pollutants breaching an air quality standard or causing injury to vegetation.

The word 'significant' has exercised the minds of many people in many organizations seeking to satisfy the searching eyes of assessors. Considering issues such as the environmental concerns, the scale of the impact, the severity of the impact and the probability of occurrence can facilitate evaluation. Modelling is one tool to support these activities. It also contributes to the emergency preparedness requirement of these environmental management system standards and the similar occupational health and safety management system (OHSAS 18001).

Public Information

Increasingly, air quality information is made available for the public in many countries, and is of particular interest to those who suffer from respiratory disorders that may be aggravated by poor air quality. Detailed regional forecasts of air quality for inclusion in weather bulletins are part of the commitment by the UK government to provide the public with up-to-date information on air quality. The forecasts cover ozone, nitrogen dioxide, sulphur dioxide, carbon monoxide and particles such as PM_{10}. The air pollution band (Table 1.1) for each of the pollutants is predicted separately for each of ten regions of the UK (Scotland, N Ireland, NW England, NE England, Wales, the Midlands, SE England, East Anglia, SW England and London). The use of banding effectively reports data in terms of air pollution rather than air quality and reflects the view that air pollution occurs when air quality deteriorates.

Other countries also use public information systems, and Figure 1.2 illustrates a US format.

Figure 1.2 US Logo for Public Awareness on Air Quality

Pollutant: Ozone

Today's Forecast: 130

Quality: Unhealthy for sensitive groups

Children and people with asthma are the groups most at risk

Air Quality Guide

In the UK, forecasters make use of information from a number of sources:

- On-line measured pollutant concentrations from the national automatic monitoring networks.
- Weather forecasts for the following day.
- Ozone data from selected European monitoring sites.
- 'Real time' results from trajectory modelling.
- Results from the urban pollutants box modelling.

The trajectory model is run each day; it is used to predict episodes of photochemically generated pollutants in the summer, where long range transport is an important factor in producing high UK concentrations. It uses the output of numerical weather prediction models as its input, and predicts 'back trajectories', ie how air masses have been transported for the preceding 96 hrs. The model uses a simplified chemical scheme to predict the formation of ozone as the air travels to the UK. This model also predicts concentrations of the secondary particle contribution to PM_{10}. Chemical models such as this are outside the scope of this introductory text.

A box model has been used to anticipate episodes that occur in cities during the winter months, due to the poor dispersion of nitrogen dioxide, sulphur dioxide, carbon monoxide and particles. It used weather forecasts for the next day and emissions estimates from the National Atmospheric Emissions Inventory (NAEI). The box model will be outlined in Chapter 2 and input data in the form of emission inventory data will be discussed in Chapter 4.

CHAPTER 2
TYPES OF MODEL

The objective of modelling is to use a tool that can reproduce some aspect of an environmental system with reasonable accuracy. Many physico-chemical processes can affect flow and transport, and not all of them may be well known. Models vary considerably in their complexity, and may take account of any of the following processes taking place in the atmosphere. Many of the terms introduced here will be developed in later chapters.

- Emissions – type and scale.
- Plume rise – associated with assisted draught or buoyant emissions.
- Transport (advection) by the wind.
- Vertical and/or lateral and/or horizontal diffusion caused by turbulence.
- Reflections from the ground and/or from the top of the mixing layer.
- Transformations of pollutants by physical or chemical processes.
- Gravitational settling of particulate matter.
- Dry deposition, ie uptake of substances on to soil and other surfaces.
- Wet deposition such as rain out and wash out of substances in clouds, rain, snow and fog etc.
- Topographical effects through variations in land surface and entrainment in the lee of buildings.
- Variations in diffusion with height.
- Variations in emissions and meteorology over time and distance.
- Random variations (ie stochastic processes).

Mathematical expressions can be derived to represent these atmospheric processes. Consequently, there is an enormous range of possible atmospheric dispersion models available for the prediction of air quality. However, the environmental characteristics may not be well established, and so the conceptualization is often difficult and hard to systematize.

In most practical situations, the modeller is faced with a large amount of data, some of which may not be quantitative. Hence, a number of assumptions may need to be made on the variations of certain parameters and on appropriate boundary conditions. The extent to which a specific model is suitable for evaluating the impact of a specific source depends on several factors that will be developed later. These include:

- the modelling objectives;
- the meteorological and topographical complexities of the area;
- the type of pollutants to be modelled (primary, inert, secondary, gas or particles);
- the spatial scale or size of the area including topographical features;
- the level of detail and accuracy needed for the analysis;
- the technical competence of those undertaking such simulation modelling;
- the financial resources and time available;
- the quality of the data base, ie, emissions inventory, meteorological and air quality data.

Classification of air quality models may be made on many criteria, which include scale, (spatial as well as temporal) and the mathematical treatment of atmospheric phenomena. Additional criteria will include source–receptor relationships (ie receptor oriented or source oriented) and the treatment of air pollutants (ie conservative or reactive). Since modelling of all the processes listed previously would need enormous input data and would increase the model complexity, the list serves as a check to recognize which processes are important or relevant in a specific application.

Model Types

Statistical Models

Statistical models are based on data from monitoring. Such models are invaluable in removing the inherent variability in raw data by compiling the distribution functions. Concentrations may be estimated for averaging times other than those used to collect the data, and this may be necessary to enable comparison against relevant standards. One example of a statistical model will be given in Chapter 3 for situations where it is necessary to account for different averaging times. The outputs from statistical models are limited in spatial distribution, and may be limited over time. There are the additional disadvantages that the monitoring facilities to generate the data can be extremely expensive, are time consuming and monitoring data may not exist for the location of interest. Another simple empirical model might be the correlation between degree-days (a measure of mean temperature linked to the demand for energy for heating or cooling purposes) and sulphur dioxide concentration averages. Such statistical relationships, if properly validated for each locality and intended use, comprise a valuable predictive tool. Often the objective is to look ahead. Empirical models can do this and predict the effects of a proposed activity or a hypothetical event. Statistical or empirical techniques are frequently employed in situations where incomplete scientific understanding of the physical and chemical processes or lack of the required data bases make the use of more detailed models impractical.

The Box Model

Air quality models vary in complexity, and one simple type of model is the box model. The concepts of conservation of mass and conservation of energy are fundamental to science and engineering. These fundamental laws of science allow an account to be made of the flows of materials and energy in and out of an environmental system, and so to track quantitatively pollutants as they disperse through the environment. A corollary of the law of conservation of mass is that the atoms in pollutants do not disappear, and transportation from one domain of the environment to another is a possibility.

Conservation of mass allows a balance to be made of the mass of pollutants as they move through the environment. However, the essential first step in developing a mass balance is to define the system. A boundary must be drawn around the system being analysed — in this case a box. In the context of air quality the system may be the atmosphere within a building, a global system or anything in between. Thus, the box may cover the site boundary of a factory or the air mass over a town. It is important to set the system boundary to keep the analysis of the environment within manageable proportions and in the context of the problem being investigated.

Whatever the system boundary, consider what happens to pollutants released into that system. Depending on the physical and chemical interactions, some of the pollutant may pass through and out of the system unchanged. Some may accumulate within the system, while some may undergo chemical transformation, such as when nitrogen oxides are converted to ozone by chemical reaction. Figure 2.1 summarizes these possibilities within the system.

Figure 2.1 An Input–Output System for an Air Pollution Model

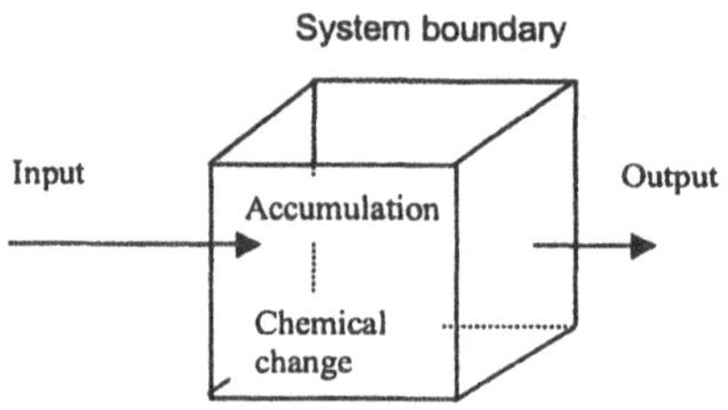

For every pollutant a mass balance equation may be written thus:

(Input rate) = (Output rate) + (Transformation rate) + (Accumulation rate)

Two simplifications can be made to the above equation. If an equilibrium condition is assumed, then the accumulation rate is set at zero. Similarly, it can often be assumed that there is no chemical reaction or radioactive decay, so that the transformation rate is also set at zero.

The model far from answers the primary question asked of a mathematical model: what will be the pollutant concentrations at any point in the air quality region, given all the data on sources and meteorological conditions? In theory, a complete answer to this question can be given only by continuously tracking the pollutants emitted from individual sources and computing the concentration of each species at every point as they are transported by the wind, spread by diffusion, mixed by turbulence and reflected or channelled by surfaces such as the ground or buildings. The basic consideration of mass continuity in fluid dynamics leads to complex equations that describe the time-varying changes in the concentration field. The problem has been likened to that of trying to describe the changing intensity of a soluble dye at every point after it is dropped into a swirling, turbulent river. However, there is an additional complication that it may be necessary to account for the chemical decay of each species with time and their reactions with each other.

So, a box model may be used for rough-order-of-magnitude assessments of a city's pollution concentration, using approximate numbers for mixing heights, wind vector and source strengths. Box models provide area-wide average concentrations and so offer a means to estimate background air quality given an emission inventory for the area. One of the most questionable assumptions of the box model concerns the uniformity of the pollutant concentration across the area under consideration. In such circumstances, multi-box models may be useful. Increasingly, data are being generated on emissions through emission inventory development. Box models can serve to check the order of magnitude of an emission inventory across a town. They are best used for daily or annual mean concentrations rather than shorter term averages. A model should also be able to account for the processes of dispersion and dilution in the atmosphere between the source and receptor. This is the role of models such as the Gaussian plume model introduced next and developed in later chapters.

Deterministic Models

By putting information on meteorology and source characteristics into a 'deterministic' model, it is possible to generate information on concentrations at any point in time or space. Such models are increasingly important for air quality management, for land use planning and for meeting a variety of regulatory requirements previously outlined in Chapter 1.

Perhaps the simplest model of a plume dispersing from a stack is where the plume expands uniformly as a cone from the height of its centre line. If the height of the centre line is H, the plume will touch the ground when it has expanded to a radius H. This simple picture is represented in Figure 2.2.

The area of the cone base is πH^2 and the amount of air passing through it is $(u\pi H^2)$ m^3s^{-1} when the wind velocity is u m s^{-1}. If the emission rate of the pollutant is Q kg s^{-1} (represented as 10^9Q µg s^{-1}), then the average pollutant concentration (C) in the expanded plume is

$$C = \frac{10^9 Q}{\pi u H^2} \mu g\ m^{-3}$$

This simple expression differs from better models of pollutant dispersal primarily in the assumption of uniform distribution of concentration around the horizontal axis, rather than the conventional Gaussian distribution to be described in Chapter 5. For the present, this simple model can be taken to represent the instantaneous conditions where the sampling time is of the order of a few minutes.

Figure 2.2 An Instantaneous Picture of a Plume Expanding as a Cone

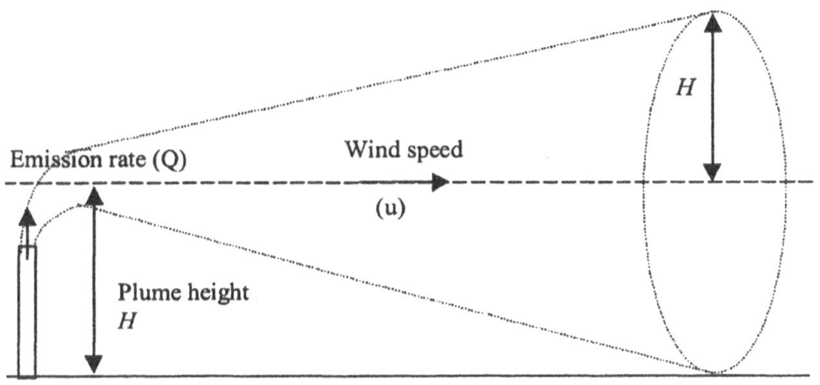

Over a long period, the wind may blow from any direction, so the cone may be oriented along any compass direction to form a cylinder. Experimentally it has been found that the maximum concentration at ground level under average wind conditions occurs at about 15 times the effective plume height from the source.* So over the long period, the cross section of the plume where it touches the ground may be at any position on the surface of a cylinder of radius 15*H* and height 2*H* as shown in Figure 2.3.

The ratio of the plume cross sectional area to that of the cylinder is

$$\frac{\pi H^2}{2 \times 15\pi H \times 2H} = \frac{1}{60}$$

This implies that the mean long term average concentration from the plume at this radius is expected to be approximately one-sixtieth of the short term concentration. In Chapter 3, a more rigorous approach will be developed. For the present, we will continue with this simple model.

* Notice the reference to effective plume height rather than stack height. A summary of early dispersion studies in the *Report of the Committee on Air Pollution* (Cmd 9322, 1954, HMSO, London) located the maximum concentration at 10 to15 times the stack height for a cold emission and 25 to 35 times the stack height for hot releases. Clearly the maxima do not always occur at the same point and the above discussion is a simplification that depends on determining a value for the effective height. This will be covered in Chapter 5.

If for this case the mass emission rate of pollutant from the source is M kg day^{-1} and taking an average wind speed as 7 m s^{-1},

$$C = \frac{10^9 M}{\pi \times 7 \times H^2 \times 60 \times 60 \times 24}$$

Converting to concentration units of mg m^{-3} this reduces to H^2 = 25M/48P. This is not unlike an expression long used by the now superseded Alkali Inspectorate, namely H^2 = 9M/20P, where P was the maximum permitted ground level concentration.

Figure 2.3 A Long Term View of a Conical Plume

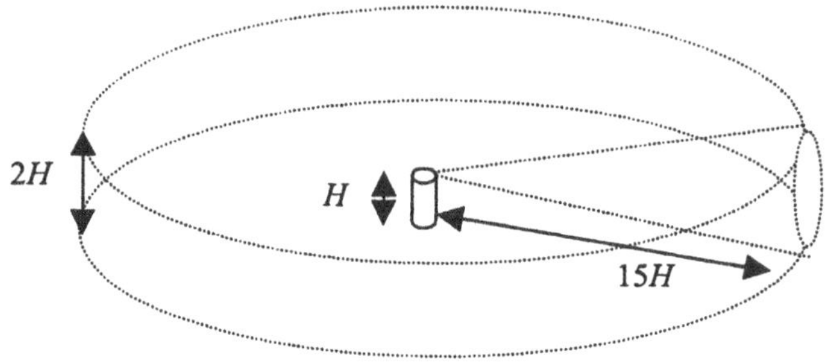

This simple treatment of a model serves to introduce some of the concepts that have developed considerably since that time. One refinement takes account of the normal or Gaussian distribution of concentration across the vertical and horizontal planes of a dispersing plume. These Gaussian models are perhaps the most widely used techniques for estimating the impact of non-reactive pollutants and will be covered from Chapter 5 onwards. Another feature concerns 'the maximum concentration at ground level'. In reality, there is not a single point; the maxima for different averaging times occur at different points and different times. This is illustrated in the edited extract from the list file created by one of the latest generations of computer dispersion models, AERMOD. This model will be discussed later. For the present it is sufficient to note that the maximum for 1 hr and for 24 hr averages occur at different locations as is evident on comparing the cartesian coordinates identified as XR and YR as shown below.

```
*** AERMOD – VERSION 99211 ***

                    *** THE SUMMARY OF HIGHEST 1-HR RESULTS ***

                    ** CONC OF SO2    IN MICROGRAMS/M**3                    **

                              DATE                          NETWORK
GROUP ID          AVERAGE CONC  (YYMMDDHH)  RECEPTOR  (XR, YR, ZELEV, ZFLAG)  OF TYPE  GRID-ID
-----------------------------------------------------------------------------------------------
ALL   2ND HIGH VALUE IS 18.84283   ON   99100513:   AT   ( 0.00, -300.00, 0.00, 0.00)   GC   GRD1

                    *** THE SUMMARY OF HIGHEST 24-HR RESULTS ***

                 ** CONC OF SO2    IN MICROGRAMS/M**3                  **

                              DATE                          NETWORK
GROUP ID          AVERAGE CONC  (YYMMDDHH)  RECEPTOR  (XR, YR, ZELEV, ZFLAG)  OF TYPE  GRID-ID
-----------------------------------------------------------------------------------------------
ALL   2ND HIGH VALUE IS 6.57062m   ON   99011724:   AT   ( 300.00, 600.00, 0.00, 0.00)   GC   GRD1
```

Computational fluid dynamics (CFD) models are the most detailed and complex dispersion models. They solve the fully three-dimensional equations of motion, where the influence of turbulence is modelled by means of turbulent eddy viscosities. These viscosities are normally functions of the turbulent kinetic energy. Additional partial differential equations are needed to describe these parameters. However, the computer requirements for this kind of model are often demanding.

Roll Back Models

The roll back model is another simple approach to predicting air quality, but will only be mentioned briefly. The roll back method uses a very simple empirical model that assumes that the average ambient concentration at any location is linearly proportional to the rate of emission. Accordingly, if emissions decrease, concentrations will decrease in direct proportion and vice versa when the substance is naturally present:

$$c = \frac{c_0 q}{q_0} = \frac{(c_0 - c_b)q + c_b}{q_0}$$

Where:

q_0 = original emission rate
c_0 = original ambient concentration (predicted)
c_b = natural background concentration
q = final emission rate
c = final ambient concentration

Roll back models assume that the receiving environment will continue to behave in the same way under the influence of new releases and other changes in the environment. The models can only be used where there are existing emissions and detectable ambient concentrations of the pollutant. Natural sources of emissions may need to be taken into account in establishing existing emissions, or else the background concentration must be allowed for in the model. There may be uncertainty over the contribution of more distant sources of emissions to existing emission levels in a particular location, and therefore, about whether these sources should be taken into account. Thus data on existing levels of emission and existing concentrations are needed. If not already available, existing emissions could be calculated using emission factors. Existing ambient concentrations may need to be measured.

Roll back models provide a simple basis for comparing alternatives on the basis that the conditions are similar. They assume linearity and take no account of differences in conditions in the receiving environment at alternative locations. They are most useful for preliminary predictions involving single sources and for comparing clearly distinguishable alternatives. They are particularly useful in preliminary assessments for planning and policy decisions that will have a direct impact on emission levels (for example fuel policies, emission control policies, regional development plans). With source apportionment studies they may help in devising action plans to improve air quality.

Physical Modelling in a Wind Tunnel

Physical modelling involves the use of wind tunnel or other fluid modelling facilities. In wind tunnel experiments the physical and/or chemical processes affecting the behaviour of pollutants released into air are simulated at a reduced scale. A scale model of the source and the topography of the receiving area (buildings, surface roughness, heights etc) is placed in a wind tunnel. Different atmospheric conditions can be simulated, including wind speed profiles and wind direction, and in a few cases atmospheric stability and boundary layer conditions. A typical wind tunnel might be tens of metres long and up to 10 m square in cross section. Rotating the model of the source and the surrounding environment within the tunnel simulates changes in wind direction. Releasing tracers simulates emissions. Instantaneous and average concentrations can be measured at locations of interest around the model site using automatic air samplers. If emissions are simulated visibly using oil fogs, photographs can provide a rapid visual presentation of the behaviour of emissions. This class of modelling is a complex process requiring a high level of technical expertise, as well as access to the necessary facilities. Facilities are available for use at various centres, but each is unique and specialists from the appropriate organization must undertake the development of models and the planning and execution of experiments. Studies may be time consuming and relatively expensive. Nevertheless, physical modelling may be useful for complex flow situations such as building, terrain or stack down-wash conditions, plume impact on elevated terrain, diffusion in an urban environment or diffusion in complex terrain. It is particularly useful for situations where a source or groups of sources in a geographical area are limited to a few square kilometres. If physical modelling is available and its applicability demonstrated it might be the best technique.

Wind tunnels can be used at different scales down to about 1/1000, and can be used to model single point emissions, roads, urban areas or topographic regions. The technique is of particular value for situations of complex source configuration or topography where mathematical modelling gives unsatisfactory results. In particular, it can be used for studying pollutant entrainment and down-wash in built up areas. It can also be used for dense gas dispersion modelling. Depending on the scale of the model and the accuracy of the predictions required, considerable input data may be required on the topography of the source and receiving environment, meteorological conditions in the study area and emission data. The accuracy of any application will depend, in particular, on how effectively atmospheric mixing layer conditions, upstream flow conditions (including topography) and plume rise are simulated. Only a relatively narrow range of meteorological conditions (usually only neutral stability) can be simulated in most wind tunnels. Special features are required to simulate stable and unstable conditions and the change of wind direction with altitude (cross wind shear), which may not always be available. Wind tunnels cannot be used for simulating calm conditions. Further discussion of physical modelling is beyond the scope of this book.

Mathematical Model Development

Early dispersion modelling focused on short range problems. In part, this was due to the lack of information about long range atmospheric chemistry. In addition, the relatively low height of stacks made pollutants most noticeable only a few kilometres downwind. Intermediate and long range transport processes have received increasing attention in recent years, especially due to the environmental problems such as acidic deposition, visibility degradation, and the ensuing, more stringent national and international environmental legislation. In addition, incidents such as the Chernobyl accident strongly enhanced the interest in long range studies. In Europe, where the observed acidification of Scandinavian rivers and lakes was the major starting point for the research in this field, many studies have been performed to derive long range numerical simulation models that better fit available air quality and meteorological measurements. Among the first studies were the Organisation for Economic Co-operation and Development (OECD) programme on Long-range Transport of Air Pollutants and the Co-operative Programme for Monitoring and Evaluation of the Long-range Transmission of Air Pollutants in Europe. Similar interest in long range dispersion grew in the US driven by the acid deposition problems in the NE United States and Canada. Although the same physical laws apply to long range air pollution problems as short range ones, different treatment is needed. This is because of the large time scale of long range transportation and factors such as deposition, chemistry and terrain complexities. Long range models and chemical models are too complex to consider in this introductory text. Brief mention will be made of particle deposition.

There are different levels of sophistication in models. The first level consists of general, relatively simple estimation techniques that provide conservative estimates of the air quality impact of a specific source, or source category. These simple models, often called screening models, serve to eliminate the need for further, more detailed modelling for sources that clearly will not contribute to ambient concentrations in excess of ambient air quality standards, nor cause a nuisance. The second level consists of those analytical techniques that provide a more detailed treatment of physical and chemical atmospheric processes, require more detailed and precise input data, and provide more specialized concentration estimates. As a result they provide a more refined and, at least theoretically, a more accurate estimate of source impact and the effectiveness of control strategies. These are referred to as refined models. Both levels of modelling will be outlined in later chapters. Appropriate data should be available before any attempt is made to apply a model. One that requires detailed, precise input data should not be used when such data are not available. However, if the data are adequate, a model that considers the spatial and temporal variations in emissions and meteorology in more detail will be better able to evaluate the source impact and to distinguish the effects of various control strategies.

Air quality models have been applied with the most accuracy, or the least degree of uncertainty, to simulations of long term averages in areas with relatively simple topography. Areas subject to major topographical influences experience meteorological complexities that are extremely difficult to simulate. Although models are available for such circumstances, they are frequently site specific and resource intensive. In the absence of a model capable of simulating such complexities, only a preliminary approximation may be feasible until better models and databases become available. The latest generation of mathematical models take account of more detailed characteristics of the atmosphere than earlier models and allow for point, area and line sources to be dealt with appropriately.

Conclusion

Modelling requires many assumptions, and there is consequently an inherent uncertainty in the predictions. However, modelling finds growing use in meeting regulatory requirements, in land use planning and as a cost-effective tool to complement monitoring. Whatever the purpose, a model must be fit for that purpose, and the modelling must be carried out in a technically competent manner. The aim of the following chapters is to give an understanding of some background to atmospheric modelling, to provide some of the skills to carry out such activities and to help the understanding of the results of modelling for risk assessment.

First, it is essential to have an appreciation of atmospheric behaviour. Meteorological models have classically concentrated attention on regions of the atmosphere in which weather systems operate. Consequently, the use of models requires an understanding of these processes and a means of representing them. This is especially important for predicting air quality; the atmospheric features that affect air quality tend to be strongly moderated by the planetary boundary layer. This is defined as the lower part of the atmosphere between the surface and the level of the geostrophic wind (ie, the level above ground where the frictional effects from the ground are small). It is that part of the atmosphere in contact with the surface within which surface-generated convection and mixing takes place. It is also the air that we breathe. Atmospheric characteristics are the topic of Chapter 3.

CHAPTER 3
UNDERSTANDING THE ATMOSPHERE

There is no complete theory of the dispersion of pollutants in the atmosphere. Instead, there are many theories, partly empirical and relatively little comprehensive data by which to choose. Nevertheless, there are many useful formulae, many of which give at least an insight into the problem. Modelling air pollutant dispersion is akin to weather meteorology and forecasting, and shares some of the uncertainties. The aim of this chapter is to explore our understanding of the atmosphere. It will refer briefly to air quality in terms of chemical composition, and hence its potential effects as a nuisance, on health, and on other aspects of environmental harm. In particular, the significance of how concentrations are expressed will be discussed in relation to aspects of statistical modelling. It will then consider its physical behaviour and define some of the input parameters needed to predict and model the transportation of pollutants. Inevitably, some theory is included. This is justified as the latest generation of dispersion models attempt to take account of the complex mathematics that describe variations of atmospheric behaviour with height.

Air Quality

While seemingly vast, the atmosphere accounts for only about 1% of the diameter of the Earth. It is also continuous and so may be contaminated by activities perhaps hundreds or even thousands of miles away. This contamination is what is called air pollution. The WHO has defined air pollution as:

> The presence in the outdoor atmosphere of one or more contaminants, such as dust, fumes, gas, mist, odour, smoke or vapour, in quantities or characteristics, and of duration such as to be injurious to human, plant or animal life or to property, or which unreasonably interferes with the comfortable enjoyment of life and property.

While this is a lengthy definition it ignores some important aspects. For example, it does not acknowledge that over 70% of our time is typically spent indoors.

Our understanding of the atmosphere has been enhanced considerably by the significant expansion in air quality monitoring. For example, in the UK there are many national networks in operation in addition to local initiatives for specific purposes. These networks provide data on air quality that are publicly available in conventional and Internet publications, and some information sources are suggested in Appendix 5.

The Effect of Averaging Time

Many of the national networks use continuously recording instruments, and typical output from these shows that air quality varies continuously. Modern sensitive instruments of the direct reading type may give instantaneous values, while with other techniques, enough sample over a period of time has to be collected in order to have enough to detect. An example is in measuring pollutants known as TOMPS (toxic organic micro-pollutants). These are determined as three-month bulk samples because the low concentrations demand long sampling times in order to collect sufficient material for analysis. Long term average values may also reflect exposure over a long period. Effectively, collecting a sample over a given time period is an averaging technique, and averaging times differ in the various national networks and in the specifications in air quality standards. An average simplifies data for interpretation, but it is essential to be aware of the consequences. This importance applies equally to understanding the results of modelling and their implications such as the potential for health effects. The time taken for exposure to a pollutant to cause adverse health effects varies from pollutant to pollutant. Some, such as sulphur dioxide, can have an effect after as little as 15 minutes, while others, such as lead, have a longer term cumulative effect. The times over which concentrations should be averaged therefore should vary to reflect this. Evidence suggests that some pollutants, for example, nitrogen dioxide, have effects after short term exposure, but that longer term exposure may also result in an adverse impact on health. In such cases, averages over different time periods can reflect the different effects. Such different averages are reflected in air quality standards as the data in Appendix 1 demonstrate.

Consider a typical variation in the concentration of an air component. Figure 3.1 represents the variation in concentrations of nitrogen dioxide in December 1991 in London. The peak hourly concentrations observed were unprecedented, reaching a maximum of 423 ppb (808 $\mu g.m^{-3}$) during a period of high pressure with low wind speeds and widespread fog. These hourly values are shown by the continuous trace in the Figure. The broken lines represent the daily average values and the horizontal lines show averages over longer periods. They indicate how increasing the averaging time leads to lower concentration values. Thus, the 8-day average is 100 ppb and this value conceals the significance of the peaks that contribute to the average. Clearly, concentration and averaging time are interlinked.

Figure 3.1 Air Quality Variations and Averaging Time: the Episode of December 1991

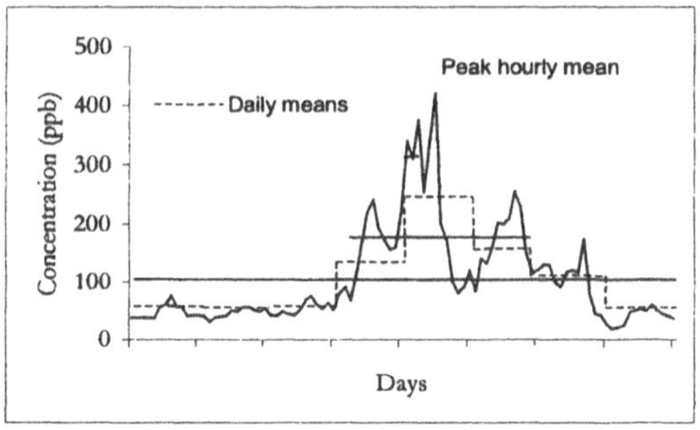

The peaks and troughs shown in Figure 3.1 show how air quality varies with time. This variation is caused by many factors, broadly grouped into human (or source) and weather effects (Figure 3.2).

Figure 3.2 Multiple Influences on Air Quality

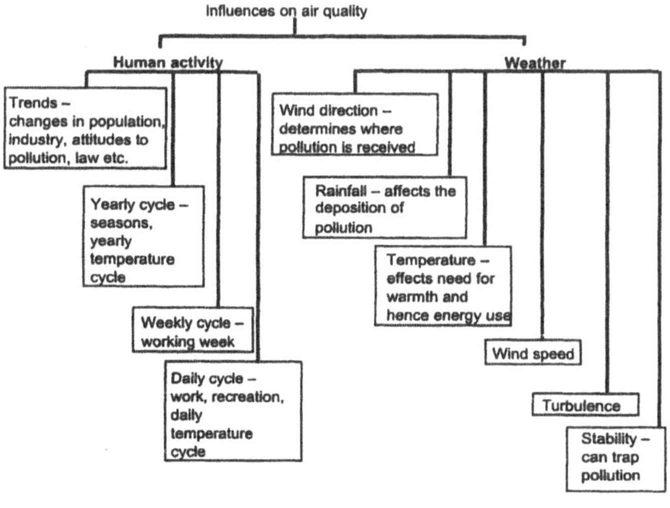

Human activities are influenced somewhat by time of day, as in the diurnal patterns of traffic flow influencing air quality. Weather patterns also have an influence. This is especially so where emissions occur through the demand for energy for lighting and heating buildings. All of these generate air pollutants, but the greatest influence of the weather is on the transportation and removal of pollution. While wind and rain are removal mechanisms, certain conditions of atmospheric 'stability' may trap pollution, as was the case in December 1991. Figure 3.6 later shows the weather map for that time. Understanding the links between air quality and weather is useful in predicting conditions under which air pollution episodes may occur, and these will be outlined later in this chapter.

Averaging over a specified period may smooth out some of the variations of air quality in Figure 3.1. This shows the effect of averaging over daily periods to give the broken lines as discussed previously. It is evident that averaging conceals peaks that may result from source peculiarities, unusual meteorological conditions or other reasons. If peak values are important, it is essential that averaging does not mask them. Similarly, modelling should generate information about concentrations using appropriate averaging time.

Raw data may be of instantaneous values or they may be averages over a particular period. This time usually depends on the time required to collect sufficient material for analysis, or may be governed by the constraints on generating a manageable set of data. Often, the averaging time ranges from several minutes to 1 hr, 24 hrs or even 1 month or more. Old measuring techniques now obsolete, such as lead dioxide candles for measuring sulphur dioxide, sampled for a month. Such long averaging times are of limited use in air quality management since the impact of events causing pollution incidents is lost in the data, as Figure 3.1 implies.

A set of raw data makes little sense at first inspection. The information becomes much clearer by plotting graphs or time sequences such as that in Figure 3.1. Such presentation techniques reveal outlying data points, regular patterns such as daily, weekly or seasonal patterns, and show up where data are missing. Episodes resulting from incidents or unusual weather conditions also show up more clearly from a graphical presentation

Statistical Summaries

The complicated pattern of data in Figure 3.3 was taken from a station generating daily averages, but the pattern resembles information generated by a modern automated station giving continuous data. The mass of data can be overwhelming. A statistical summary is useful to describe such data.

Figure 3.3 Air Monitoring Data for Sulphur Dioxide ($\mu g.m^{-3}$)

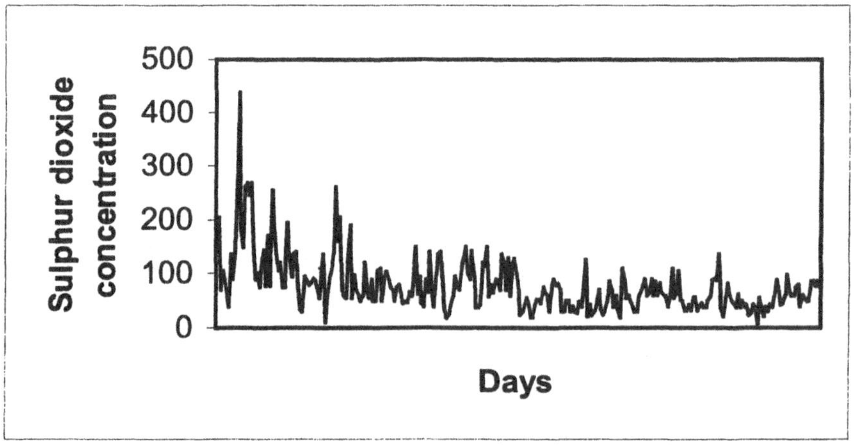

The first step is to rank or arrange the data from the lowest to the highest value. Where there are many results, it may be preferable to group the data as shown in Table 3.1, which uses the data collected over 24 hr averaging times presented in Figure 3.3.

Table 3.1 Frequency Distribution of Air Quality Data

Concentration interval	No of days	Relative frequency (%)
0—25	13	3.6
26—50	85	23.2
51—75	108	29.5
76—100	76	20.8
101—125	26	7.1
126—150	28	7.7
151—175	10	2.7
176—200	6	1.6
200+	13	3.6

The difference between the minimum and maximum value is the range of the data set. The middle value (or the arithmetic mean of the two middle values if the number of measurements is even) is the median. By dividing the set into four, quartiles are obtained, while ten divisions give deciles and ultimately one may define percentiles. The median is therefore the 50th percentile. Air quality standards in the EU and UK are specified using percentiles either explicitly or implicitly. For example, the EU requirement for sulphur dioxide is that 125mg m^{-3} is not to be exceeded more than 3 times a year. This equates to the 99th percentile.

Another way of examining frequency distributions is to group the data and to generate histograms — ie graphical representations of the frequency distributions. This analysis starts by arranging the set of data in order of magnitude and distributing them into classes or categories. Commonly, a uniform range is chosen that is suitable for all ranges likely to be encountered for the pollutant under study. The relative or percentage frequency distribution is obtained by expressing the frequencies as percentages of the total number of observations. The data from Table 3.1 have been treated in this way to derive Table 3.2.

Table 3.2 Cumulative Frequency Distribution of Air Quality Data

Daily mean (μg m^{-3})	Cumulative frequency	% Cumulative frequency
< 25	13	3.6
< 50	98	26.8
< 75	206	56.4
< 100	282	77.3
< 125	308	84.4
< 150	336	92.1
< 175	346	94.3
< 200	352	96.4
< 300	365	100

The histogram in Figure 3.4 was derived from the data in Figure 3.3, but with more concentration ranges than shown in Table 3.2. The widths of the rectangles in the histogram are the class interval and the heights represent the frequencies. Different intervals have been used from those in Table 3.2. By fitting a continuous curve to the histogram, a frequency curve is produced. As the width of the rectangles become smaller (and so their number becomes greater), so the frequency curve becomes smoother. The total frequency of values below the upper limit of a given class interval is known as the cumulative frequency up to and including that class. A cumulative frequency curve is included in Figure 3.4. A useful feature of the cumulative frequency curve is that the percentage of measurements below any specified value can easily be shown.

The Log-normal Distribution

Figure 3.4 shows that air quality data tend to be positively skewed (ie have a tail to larger values) when plotted as a histogram. This asymmetry is inconvenient compared with the more familiar, normal distribution, with its symmetrical bell shape.

Figure 3.4 Histogram of Air Quality Data

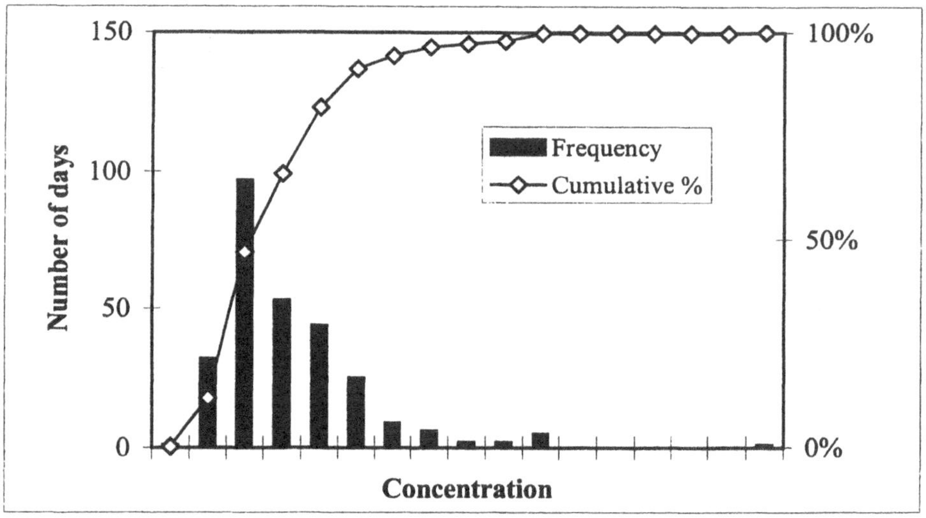

A logarithmic transformation can change an asymmetrical into a symmetrical histogram rather like the normal distribution. Hence, the original set of air quality data is called log-normally distributed. A log-transformation of the data has been done to derive Figure 3.5. The bell-shaped curve characteristic of the normal or Gaussian distribution is now apparent.

Figure 3.5 Histogram and Frequency Curve for Air Quality Data Using Log-transformed Concentrations

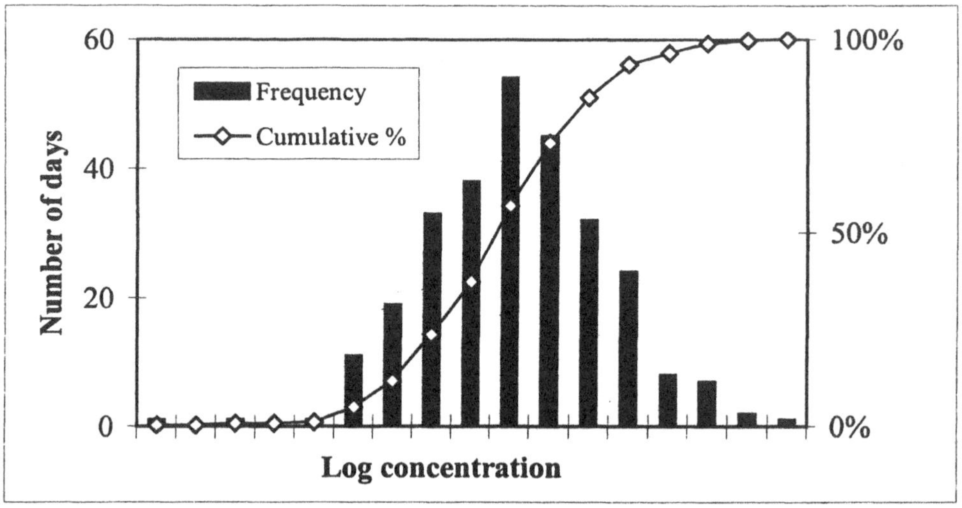

Just as the mean and standard deviation can represent normally distributed data, it is possible to describe log-normal data by two simple representative values — the geometric mean and the standard geometric deviation. The geometric mean concentration (m_g) for a log-normal distribution falls at the median or 50th percentile point. The slope of the line gives the standard geometric deviation.

However, in a true log-normal distribution, one standard geometric deviation (s_g) from the geometric mean (m_g) corresponds to the 16th or 84th percentile. This provides a simple way of estimating the standard geometric deviation — ie the slope of the line:

$$S_g = \frac{16^{th}\, percentile}{50^{th}\, percentile}$$

Various statistical distributions have been applied to air quality data, but the log-normal distribution as described above has been the most popular. In fact, it is followed more or less closely by many sets of environmental data, especially those characterized by the fact that compared against the mean value, most observations are small and few are large. Many types of data other than air quality follow this distribution, and include trace metals in environmental samples, water quality, the heights of floods in a river and annual salary levels.

The equation that describes the log-normal distribution of a concentration (c) is:

$$p(c) = \frac{1}{\sqrt{2\pi}\, \ln s_g}\, e^{-\frac{\left(\ln c - \ln m_g\right)}{2 \ln^2 s_g}}$$

Just as there are many normal distributions, so there are many log-normal distributions, depending on the values of s_g and m_g. All are skewed to the right, the degree of skewness depending on the value of the variance m_g^2. If the value of m_g^2 is small, so is the extent of skewness, and the frequency distribution tends towards a normal distribution.

The wide use of the log-normal distribution in analysing air quality data is mainly due to:

- its adequate description of the general behaviour of air quality concentrations;
- the convenience of visualization on log-probability paper; and
- its usefulness as a model in dealing with problems with air quality data.

An example will now be given.

Dealing With the Effect of Averaging Time

It has been noted that air quality data are generally reported as time-averaged values, with the averaging time being up to one year. Observations have shown that air component concentrations follow the log-normal distribution regardless of the averaging time used. This forms the basis of mathematical models for converting data obtained over one averaging time to any other averaging time.* There is a general concept that the longer the averaging time, the lower the value of the maximum concentration. The horizontal lines in Figure 3.1 illustrated this.

Studies of air quality data have produced a model that makes it possible to calculate the geometric mean for one averaging time from information from another averaging time. One equation for this transformation is:

$$m_{gb} = m_{ga} e^{\left[0.5(1-\nu)\ln^2 s_{ga}\right]}$$

where:

m_{ga} is the median (geometric mean) for an averaging time 'a';

m_{gb} is the value for an averaging time 'b' and

* Larsen R I, *A mathematical model for relating air quality measurements to air quality standards*, US Environmental Protection Agency publication AP-89, November 1971.

s_{ga} is the standard geometric deviation at averaging time a.

The value for v is given by:

$$v = \frac{\ln\left(t_{tot}/t_b\right)}{\ln\left(t_{tot}/t_a\right)}$$

where t_{tot} is the total averaging time (usually hours in a year), and t_a and t_b are the averaging times.

As an example consider that the geometric mean of 1hr average concentrations for sulphur dioxide is 0.042 ppm and the standard geometric deviation is 1.96. The geometric mean for daily average concentrations is the desired statistic. In this example t_{tot} = 8760; t_a = 1 and t_b = 24

Substituting in the equations:

$$v = \frac{\ln\left(t_{tot}/t_b\right)}{\ln\left(t_{tot}/t_a\right)} = \frac{\ln\left(8760/24\right)}{\ln\left(8760/1\right)} = 0.65$$

so

$$m_{gday} = m_{ghour} e^{\left[0.5(1-v)\ln^2 s_{ghour}\right]} = 0.042 e^{\left[0.5(1-0.65)\ln^2 1.96\right]}$$

$$= 0.045$$

Notice that these concentrations are geometric means and not maxima to which the inverse relationship applies as noted above.

Turner* observed that the effects of sampling time are 'exceedingly complex' and tend to decrease with longer sampling time as a result of increased meandering of the wind direction. This effect is illustrated later in Figure 5.3. Turner gave typical variations as set out in Table 3.3.

Table 3.3 Variation of Calculated Concentration with Sampling Time

Sampling time	Ratio of calculated concentration to 3 min concentration
3 min	1
15 min	0.82
1 hr	0.61
3 hr	0.51
24 hr	0.36

Source: Turner (1994)

According to Turner, the best estimate was available from:

$$X_s = X_k\left(\frac{t_k}{t_s}\right)^p$$

In this expression, X_s is the desired concentration estimate for the sampling time, t_s while X_k is the concentration estimate for the shorter sampling time, t_s. The value of p is typically 0.17 to 0.2.

Other mathematical models have been published.† While such equations can be used for these concentration conversions, a 'short-cut' route has been proposed. Table 3.4 has been compiled from various published sources.

* Turner D B, *Workbook on Atmospheric Dispersion Estimates*, US Public Health Service Publication No. 26, 1970; Second Edition, Lewis Publishers, 1994.

† Pratt M & Dalton H, 2000, 'A method for calculating short period concentration statistics from annual mean concentrations', *Clean Air*, 30 (3), 88-93.

Table 3.4 Ratio of Expected Annual Maxima of Concentrations for Various Averaging Times and Selected Standard Geometric Deviations

Standard geometric deviation for averaging times								Ratio of annual maximum concentration to mean concentration for averaging times of:							
1s	5min	1hr	3hr	8hr	1day	1mon	1y	1s	5min	1hr	3hr	8hr	1day	1mon	1y
1.0	1.0	1.0	1.0	1.0	1.0	1.0	1.0	1.0	1.0	1.0	1.0	1.0	1.0	1.0	1.
1.14	1.11	1.10	1.09	1.09	1.08	1.05	1.04	2.04	1.59	1.43	1.37	1.31	1.25	1.08	1.0
2.16	1.88	1.75	1.69	1.63	1.57	1.34	1.24	51.78	13.11	7.21	5.55	4.40	3.40	1.55	1.2
2.69	2.25	2.05	1.96	1.88	1.78	1.46	1.32	141.88	25.24	11.91	8.58	6.40	4.64	1.73	1.3
3.93	3.07	2.70	2.54	2.39	2.23	1.68	1.47	732.07	73.17	26.87	17.35	11.74	7.64	2.05	1.4
4.66	3.52	3.05	2.84	2.66	2.46	1.79	1.54	1445.27	113.64	37.60	23.20	15.07	9.38	2.20	1.5
5.08	3.78	3.25	3.02	2.81	2.59	1.85	1.58	2037.07	141.87	44.53	26.85	17.09	10.40	2.27	1.5
5.52	4.05	3.45	3.19	2.97	2.71	1.91	1.61	2793.31	173.97	52.01	30.71	19.17	11.44	2.35	1.5
6.42	4.58	3.85	3.54	3.27	2.96	2.02	1.69	4901.25	250.06	68.55	38.97	23.53	13.53	2.48	1.6

An example referring to the boxes in Table 3.3 will serve to illustrate its use. If the standard geometric deviation of 24 hr average values measured at a site is 1.78, the highest annual 24 hr average is expected to be 4.64 times the annual arithmetic mean. Similarly, the highest annual 1 hr average is expected to be 11.91 times greater than the annual mean. Notice that these data follow the pattern of an increase in maximum concentration as the averaging time becomes shorter. This was what Figure 3.1 implied.

The following guidelines (Table 3.5) were suggested by the former HM Inspectorate of Pollution (HMIP) as approximate conversion factors for correcting data from one averaging period to another in integrated pollution control (IPC) assessments. It was pointed out that these values apply to estimated concentrations downwind of a single elevated source. For ambient air quality, for which many sources are involved, different relationships will exist.

Table 3.5 Relationships Between Different Averaging Times from Elevated Sources

Sample reference period	Approximate ratio to 1 hr reference period
3 min	3
30 min	1.5
1 hr	1
8 hr	0.25
24 hr	0.2

More recently, a factor of 1.34 was assumed for converting 1 hr mean concentrations predicted by the ADMS model to the 15 min means advocated in local air quality guidance.* This is comparable to the relationship implied in Table 3.5.

Another example comes from guidelines for SCREEN, a simple air quality model to be introduced in Chapter 6. To obtain the estimated maximum concentration for a 3, 8 or 24 hr averaging time, the 1-hr maximum should be multiplied by the factor indicated in Table 3.6.

Table 3.6 Conversion Factors for Averaging Times for the SCREEN Model†

Averaging Time	3 hr	8 hr	24 hr	Annual
Multiply 1hr value by	0.9 (±0.1)	0.7 (±0.2)	0.4 (±0.2)	0.08 (±0.02)

*Source: EPA**

While there are clear differences between the data in Tables 3.3 to 3.6, the trends are broadly similar. However, it must be stressed again that caution is necessary when comparing air quality data against

* Vawda Y, Moorcroft J S, Khandelwal P and Whall C, 2001,' Sulphur Dioxide Emissions from Shell Boilers – Supplementary Assistance on Stack Height Determination', *Clean Air*, 31, 89–95.

† *Screening Procedures for Estimating the Air Quality Impact of Stationary Sources*, Revised (EPA-454/R-92-019) US Environmental Protection Agency, Office of Air and Radiation, Office of Air Quality Planning and Standards, Research Triangle Park, North Carolina 27711, October 1992.

one another and against standards. The averaging time is critical. Moreover, in the US, the conversion from long term to short term concentration averages by any transformation technique is not acceptable in regulatory applications.

The factors just outlined have important implications for interpreting air quality data and for interpreting the predictions from models to be covered in later chapters. The averaging time must reflect the environmental impact under consideration. So, for example, the response to odours occurs after exposure of as short as one second. Measurement or modelling air quality over, say, 1 hr, will not reflect that the value over one second may be considerably greater. The importance of this must not be underestimated.* Evidence at a public inquiry concerning odour nuisance included data based on 15-minute averages. The counter argument that one-second's exposure may be to concentrations many times greater won the day! Similarly, any data must reflect the averaging time specified in air quality standards. Unlike routine meteorological operational products, air concentration maps represent an average over a period, rather than a snapshot valid at a particular time. Therefore, when interpreting this output, the illustrated plumes will always be larger than the actual plumes as might be seen from an aircraft or satellite picture. The larger size is due to advection (wind speed) and dispersion processes that occur over the averaging period. The advection or the time the plume spends over a particular point, and the horizontal and vertical mixing will affect the concentration levels. These are the factors to be considered when comparing different model products. Is the advection in the same direction? Is the advection in a different direction because one model may mix the pollutant to higher levels above ground?

Physical Characteristics of the Atmosphere

Models for estimating the impact of a source of pollution on air quality draw on information listed in Chapter 2 on the meteorological characteristics of the atmosphere. The principal characteristics are:

- turbulence in the atmosphere described in terms of convection and stability;
- wind speed and direction;
- mixing height; and
- temperature.

It is therefore necessary to consider each of these in more detail. Before doing so it is useful to consider some general features of meteorology in order to clarify the relevance of these terms.

Reading a Weather Map

A typical weather chart on television or in a newspaper shows various types of symbols representing different meteorological features. Figure 3.6 shows the chart relating to the episode from Figure 3.1. Thin lines represent isobars, which indicate areas of equal atmospheric pressure. 'HIGH' or 'H' means a high-pressure area or anticyclone. In summer, this usually makes the weather warm and sunny, while in winter it often causes cold weather and fogs. A ridge is a high-pressure region that is not at the centre of an anticyclone. Anticyclones may also be centred over or near a region, and give rise to fair, stable conditions because of the clear skies and light winds. They have been associated with several major pollution episodes in the past, including that from 1991 represented in Figures 3.1 and 3.6.

'LOW' or 'L' means a low-pressure region, depression or cyclone, which is usually associated with wind and rain. Flow is outward, downward and clockwise around a high-pressure region, and

* Hall, D J and Kukudia, V, 1994, 'Approaches to the Calculation of Discharge Stack Heights for Odour Control', *Clean Air*, 24 (2) 74 – 92.

inward, upward and anticlockwise around a low-pressure region. A trough is an area of low pressure away from the centre of a depression.

Bold lines with marks are fronts, where air masses of different temperatures meet. Cold fronts (by convention marked with triangles) have a narrow belt of rain ahead of them, but as the front passes, brighter weather follows. Warm fronts (marked with semi-circles) show rain, but once they have passed, they are followed by wet, cloudy weather. Occluded fronts are areas where warm and cold fronts meet. Ahead of them is rain, but they are followed by dry, bright weather. Fronts provide good mixing conditions to disperse air pollution.

Figure 3.6 A Simplified Weather Map for 12 December 1991

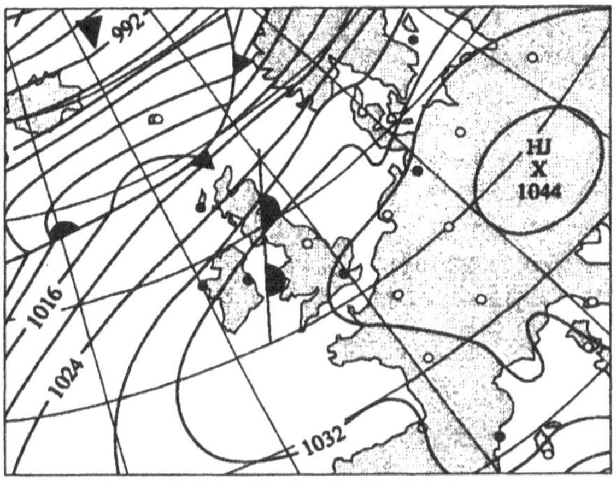

Lapse Rate

The dispersion of air pollutants through the atmosphere depends largely on the speed and direction of the winds, but the vertical mixing is influenced strongly by differences in temperature with altitude. Sunlight and the greenhouse effect warm the Earth's surface. A hypothetical bubble of air next to the surface is warmed and rises as it becomes less dense. In rising, it moves to a lower pressure region and so expands. If the bubble neither gains nor loses energy from its surroundings (called an adiabatic expansion), the energy for expansion comes from within the bubble, and so it cools. The rate at which the atmospheric temperature changes with height is called the lapse rate, and it is a major factor influencing the dispersion of pollutants and hence air quality. Basic physics can be used to determine that the rate of change of temperature with height is given by:

$$\frac{dT}{dz} = -0.0098 \text{ K m}^{-1}$$

The important result is that as a parcel of air rises through the atmosphere, theoretically its temperature will fall by 9.8°C for every 1 km rise in altitude and vice versa. This is approximately 1°C for every 100 m change in altitude. This is known as the dry adiabatic lapse rate, and its value is usually written as -1°C 100 m^{-1}.

It is important to note that if the air is laden with water vapour, as the air parcel cools, the water vapour can condense and in so doing will release its enthalpy of evaporation. Thus, a very wet parcel of air will not cool as much as a dry parcel. As the amount of water in the atmosphere is variable, there is not a single wet adiabatic lapse rate. If the air is moist, the cooling effect may allow water to condense and form clouds, with the release of enthalpy of evaporation (latent heat) reducing the cooling rate.

As the atmosphere is not quiescent, variable heating of the ground from one area to another and over time causes the lapse rate in the lowest few hundred metres of the atmosphere to show marked variations throughout the day. If the temperature at various heights in the atmosphere were measured this is known as the environmental lapse rate, and may not be the same as the dry adiabatic lapse rate.

The vertical temperature gradient, or lapse rate, is generally used as an indicator of atmospheric stability, which simply, is the tendency of the atmosphere to resist or enhance vertical motion. Figure 3.7 shows ways in which plumes discharged from stacks tend to behave under different conditions of atmospheric stability indicated by the lapse rate. The resulting effect on the ground level concentration of pollutants from the plume is also shown in the form of typical recorder traces from instruments measuring air quality.

Figure 3.7 The Influence of Environmental Lapse Rate on Plume Behaviour and Ground Level Concentrations of a Pollutant

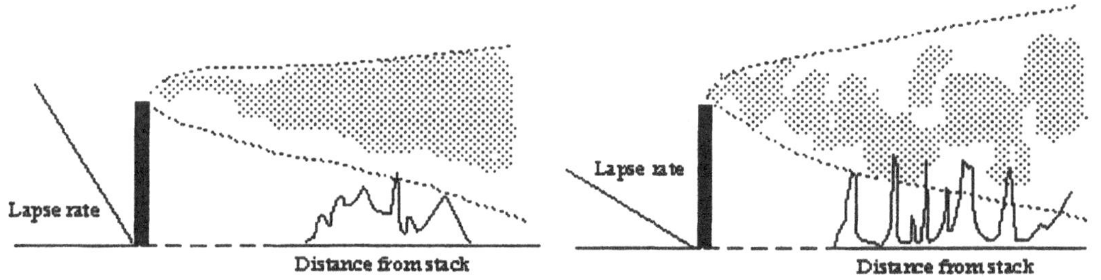

(a) coning plume **(b) looping plume**

Neutral Conditions

When the environmental lapse rate approximates to the dry adiabatic lapse rate, it is envisaged that a mass of air cools at the same rate as the environment surrounding it. Consequently, there is no tendency for a displaced mass of air to move up or down. The environmental lapse rate only approximates to the 'ideal' dry adiabatic lapse rate if there is thorough mixing of the air. This is due to factors such as winds, the spin of the earth, variable heating and cooling, and the role of water vapour. Neutral conditions are associated with overcast skies and moderate to strong winds; the clouds prevent solar radiation from reaching the surface, while the wind mixes the air and smoothes temperature differences. Such conditions are common in the UK. A plume discharged under these conditions spreads out from the stack as a steadily expanding but slightly tilted cone. Eventually the lower edge of the expanding cone or waste gases will reach the ground, although dispersion and dilution will continue downwind. Beyond the point of first contact and in a horizontal direction, the ground level concentration of a pollutant tends to rise rapidly to a maximum at progressively greater distances, before slowly declining. In the absence of other discharges, the pollution will continue to fall to the background concentration. Figure 3.7(a) represents this behaviour over open level country. The behaviour over uneven landscape and around buildings is more complicated and will be described later in the chapter. This coning plume is the basis of many atmospheric dispersion models that are widely used in air quality management.

The Superadiabatic Case

When strong solar heating occurs or when relatively cold air is transported over a much warmer surface, the temperature gradient is usually greater than the dry adiabatic lapse rate. Such conditions generally favour strong convection and turbulence in the air. This superadiabatic case creates an unstable atmosphere, which typically occurs under strong solar heating. The sun warms the ground, which in turn heats the air next to it, but the air cannot redistribute itself to accommodate the heat input. Therefore, up to 50 m of the atmosphere is unstable. While the instability is good for dispersing pollutants, it is possible for a plume to develop 'looping' characteristics, and reach the ground quite close to the source before much dilution of pollution has occurred. Figure 3.8 illustrates a looping plume. Under such circumstances high pollutant concentrations may result as shown in Figure 3.7(b). Large-scale unstable lapse rates are rare because the buoyancy forces cause the air to redistribute and form neutral conditions.

Figure 3.8 A Looping Plume

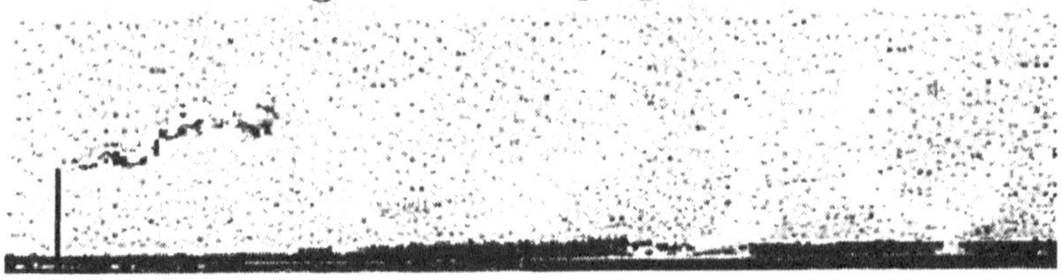

Inversion

There are many occasions when, at increasing heights above the ground, the air temperature increases. Because this is the reverse of the expected relationship between air temperature and height, it is called a temperature inversion. Such an inversion may extend from ground level or may develop at a substantial height above ground.

Figure 3.9 (a and b) shows the behaviour of stack plumes under inversion conditions, which may effectively act as a lid on the atmosphere, and prevent good dispersion. Inversions are important to studies of the dispersion of atmospheric contaminants. This is because intermixing of air layers lying above and below a 'strong' inversion is virtually excluded. The 'strength' of an inversion depends both on the minimum negative value of the temperature lapse rate and on the effective depth of the inversion layer. The 'stronger' the inversion, the greater is the 'barrier' to intermixing. If the barrier forms above a release point such as a stack, the effluent may have insufficient buoyancy to penetrate the inversion and disperse upwards. Ground level concentrations of pollutants may then be increased dramatically. These circumstances have produced major pollution episodes throughout the world. Conversely, if the inversion forms below the point of discharge, the effluent is unlikely to reach the ground near the stack and the expected pollution at ground level will be much reduced. Ideally a stack discharge rises above the inversion layer (a lofting plume) which hinders the return of pollutants to ground level.

Figure 3.9 Influence of Inversions on Ground Level Concentration

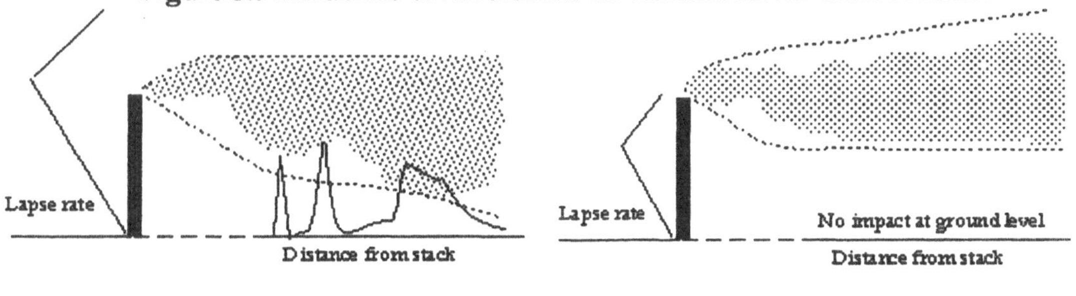

a) inversion height above plume **b) inversion height below plume**

Inversions form through cooling from below or warming from above. Low level inversions tend to form overnight when winds are light and skies are clear. After a sunny day the ground cools down as the sun declines. Without cloud cover, heat is radiated steadily into space and layers of air just above the ground are cooled by conduction. Under these conditions, mist or fog will form if the original air humidity was sufficiently high for the newly cooled air to reach its saturation point. With continued cooling, the depth of the inversion layer will gradually extend upwards to greater heights. When the sun begins to warm up the ground again after dawn, air mixing will be renewed and the inversion will normally be dispersed. These low level inversions are known as radiation inversions. A nocturnal radiation inversion commonly causes fogs, as droplets of water vapour condense from the cooled air. The presence of particulate matter (such as smoke) may serve as a nucleus for the droplets and so stabilize the fog. Such conditions gave rise to the notorious London smog of 1952, the name of which was invented in the early part of the 20th century by H A Des Voeux from SMoke and fOG.

Inversions formed in a layer of air some distance above the ground are called elevated inversions. They tend to be much more widespread and more persistent. Such inversions can be formed by a number of mechanisms, of which subsidence is noteworthy. Subsidence of a mass of air from aloft causes the air to become warmer adiabatically. This results in the warmer layer coming to rest above an original cool stagnant air mass. If clouds are present below the warm layer, radiation from their upper surfaces will accentuate the inversion. Subsidence inversions are usually associated with anticyclones or high-pressure weather systems that frequently produce clear skies. These in turn may allow ground based radiation inversions to form. Anticyclones are also associated with low wind speeds that do not aid dispersion.

Identifying Atmospheric Stability

The behaviour of a dispersing plume in the atmosphere is related to the lapse rate. The significance of the lapse rate is in terms of its impact on atmospheric stability. Simpler models use stability categories as indicators of atmospheric turbulence. The stability category at any given time will depend on static stability (related to the change in temperature with height), thermal turbulence (caused by heating of the air at ground level), and mechanical turbulence (a function of wind speed and surface roughness). The atmospheric conditions determine the distribution of air pollution impacts for a particular area.

In the summer of 1956, experimental observations in Nebraska (project Prairie Grass) resulted in Frank Pasquill suggesting categories of atmospheric stability that could easily be identified from readily available information.* These categories are very important, and despite more quantitative expressions to be described later, they remain in common use with many of the often-used models for atmospheric dispersion. The Pasquill approach uses the letters A to F to designate atmospheric stability as shown in Table 3.7.

A Qualitative Approach

The simplest approach to assessing atmospheric stability depends on knowing the surface wind speed at 10 m altitude and assessing qualitatively the amount of solar radiation (or insolation). This approach gives a rule-of-thumb method in the absence of detailed meteorological data. Table 3.8 summarizes the essential features of the method and refers back to the Pasquill stability categories defined in Table 3.7. In order to illustrate the use of Table 3.8, consider a situation in which the wind speed is 6 m s^{-1} and clear skies allow a high degree of insolation. Under these conditions, Table 3.8 suggests a stability category C.

From Table 3.8, if the wind speed was 2 m s^{-1} and there was slight insolation, this could indicate stability category B, although it could be argued that overcast conditions suggest category D, according to the footnote. This element of subjectivity in the assessment can lead to ambiguity and uncertainty. However, it is possible to apply a quantitative approach to the assessment rather than using the terms strong, moderate or slight. Table 3.9 helps in this estimation.

* Pasquill F, 1961, 'The estimation of the dispersion of windborne material', *Meteorol. Mag.* 90, (1063), 33-49.

Other methods require information on solar elevation angle, cloud cover, cloud ceiling height and wind speed. Opaque cloud cover should be used if available, otherwise total cloud cover may be used. The solar elevation angle is a function of the time of year and the time of day. Based on quantitative measures, strong insolation is regarded as exceeding 600 W m^{-2}, moderate falls within the 300—600 range while slight insolation is less than 300 W m^{-2}. Some computer models introduced in later chapters make these calculations.

Table 3.7 Understanding Atmospheric Stability Categories

Stability category	Classification	Natural phenomena	Most likely occurrence	Frequency of occurrence in central England (%)
A	extremely unstable	Strong thermal instability, bright sun	Late morning to mid afternoon in spring and summer	0.6
B	moderately unstable	Transitional period, moderate mixing	Day time transitions all year	6
C	slightly unstable	Transitional periods, slight mixing	Day time transitions all year	17
D	neutral	Strong winds, overcast, day/night transitions	Day time/cloudy; night-time/cloudy; high wind, day transition all year	60
E	slightly stable	Transitional periods, night-time moderate winds	Night time transition all year	7
F	moderately stable	Clear night time skies, very limited vertical mixing, plume fanning and meandering	Night, clear skies, light winds all year	8
G	This category was introduced later for extremely stable conditions			1.4

Table 3.8 Atmospheric Stability Categories Based on Wind Speed and Insolation

Wind Speed at 10 m ($m.s^{-1}$)	Day: Incoming Solar Radiation (Insolation)***			Night**	
	Strong	Moderate	Slight	Thinly Overcast or ≥ 4/8 Low Cloud Cover	≤ 3/8 Cloud Cover
< 2	A	A – B	B	F	F
2 – 3	A – B	B	C	E	F
3 – 5	B	B – C	C	D	E
5 – 6	C	C – D	D	D	D
> 6	C	D	D	D	D

* The neutral class (D) should be assumed for all overcast conditions during day or night.

** Night is defined as the period from 1 hr before sunset to 1 hr after sunrise.

*** Insolation categories are determined using sky cover, solar elevation information and Table 3.9.

Table 3.9 Estimation of Insolation

Sky Cover (Opaque or Total)	Solar Elevation Angle > 60°	Solar Elevation Angle ≤ 60° but > 35°	Solar Elevation Angle ≤ 35° but > 15°
4/8 or less or any amount of high, thin clouds	Strong	Moderate	Slight
5/8 to 7/8 middle clouds (2000 to 5000 m base)	Moderate	Slight	Slight
5/8 to 7/8 low clouds (less than 2000 m base)	Slight	Slight	Slight

The important meteorological parameters used by Gaussian dispersion models are the wind speed, wind direction, ambient temperature, atmospheric stability category and the mixing height. There is a general trend away from models based on Pasquill Stability towards models needing more sophisticated meteorological data. The 'new generation' of models being developed has a better representation of the physics in the atmospheric boundary layer than the 'old generation' of Pasquill type models. The reasons for this will be apparent later in this chapter. The new models require input of not only routinely measured meteorological data, but also of fundamental parameters that describe the atmospheric dispersion, such as the Monin–Obukhov length, friction velocity and boundary-layer height. The fundamental dispersion parameters can be derived from the measured meteorological parameters, but at the expense of a considerable amount of pre-processing. Some of these parameters will also be discussed later in this chapter. Dispersion models themselves will be introduced in later chapters. However, it may be noted here that in simple so called 'screening models', such as SCREEN, pre-defined combinations of meteorological conditions are examined. By comparison, refined models, such as ISC and ADMS, described later, use observed, real-time meteorological data.

It is significant to note that Table 3.8 specifies wind speed at 10 m altitude. This is meteorological convention, and in the absence of indications to the contrary, it should be assumed that this is the case. Recalling the suggestion on page 15 that height above ground influences wind speed, it will be necessary later to consider the significance of height in more detail.

Wind Direction and Speed

Wind direction has perhaps the greatest impact on air quality at a given point, for changes in wind direction have a great effect on the pollution received from a source. The importance of wind direction is such that any receptor is only at risk if situated downwind of a source of air pollution. However, wind varies in both time and position.

Wind direction is a primary measurement in air pollution studies. Data from a local weather station showed the following distribution of wind direction according to time at that point (Figure 3.10).

In this wind rose, the length of each 'arm' is proportional to the amount of time the wind blows *from* the respective direction. In a similar manner, it is possible to relate wind direction with air quality measurements by a pollution rose in which the length of each arm is drawn in proportion to the average pollutant concentration when the wind blew from that direction. So a pollution rose suggests the direction in which a source of pollution lies.

Information on wind direction and speed are required to estimate short term peak and long term average concentrations. The wind speed is used to determine plume dilution and plume rise downwind of the stack. These factors, in turn, affect the magnitude of and distance to the maximum ground level concentration.

Figure 3.10 Frequency of Wind Speed and Direction Represented as a Wind Rose (Birmingham 1999)

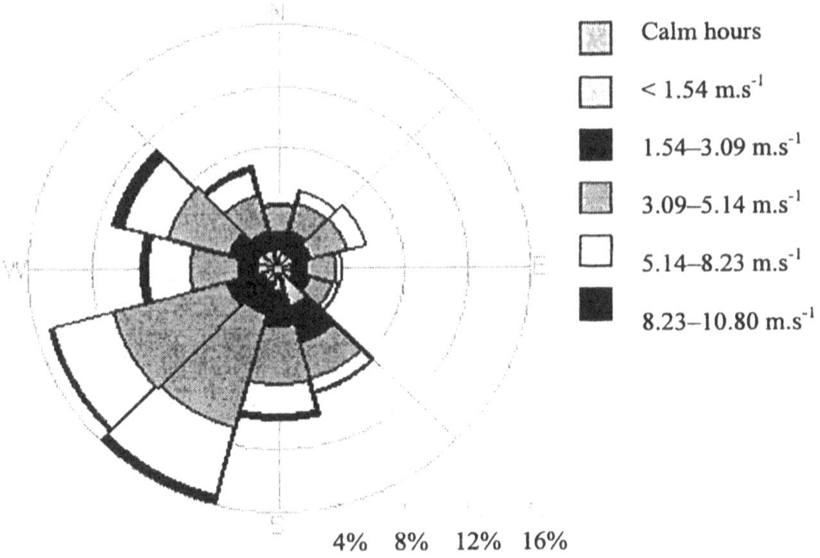

The mixing power of the air and its capacity to dilute pollutants depend very much on the speed of the wind and its gustiness. These characteristics vary widely, both from time to time and place to place. Consider gaseous effluent being emitted in a steady stream. If the wind speed is doubled, the pollutant will be dispersed into approximately twice the volume of air so that the concentration at any point is halved. After being discharged, the area over which a release passes naturally depends on the direction of the prevailing wind. It is evident, therefore, that the choice of wind speed is a critical factor in the input to a model.

It is usual for the wind speed to approximately double between 10 m and 1 km above ground; it will also change direction with height. Near the ground, objects such as stones, rocks, vegetation, buildings and hills mechanically stir the wind to form eddies. In this way, some of the energy in the wind is dissipated and the wind speed drops. The upper layers of air, now moving faster, have to slide over the lower layers and this causes interlocking, an effect that propagates eddies to progressively greater heights. Therefore, plume rise is important. The turbulence of eddies, or gusts, is conveniently described as random fluctuations of air velocity, in any direction, which are superimposed on the mean flow. Eddies are also formed during the day when one patch of ground absorbs more heat from the sun than another. Air passing over it is warmed and, becoming lighter rises through the air mass as a convection current. Such upward streams, or thermals, add to the general level of turbulence by the mixing effect at their boundaries. Once eddies are formed, they use up their energy to overcome the viscosity of the air and they become progressively smaller. At low wind speeds, turbulence would tend to disappear were it not for the fact that new eddies are constantly being formed. The stirring process, due to the unevenness of the ground and buildings, thus maintains a state of turbulence, which tends to decrease with height, up to gradient height. For convenience of description, the moving mass of air in which this process becomes established is called the atmospheric boundary layer.

This is a simple description of a common dynamic situation, but it is often profoundly modified by the way air temperature varies with height, which can have dramatic and far-reaching effects on the dispersion process. This happens for a variety of reasons such as solar heating, radiation cooling at night and variations in cloud cover, etc.

Most wind data are collected near ground level. The wind speed at height, u_s, can be estimated from the following power law equation:

$$u_s = u_1 \left[\frac{h_s}{z} \right]^p$$

where:

u_s = the wind speed (m s^{-1}) at height, h_s,
u_1 = the wind speed at a reference height, z (such as the anemometer height), and
p = the stability-related power law exponent from Table 3.10.

Table 3.10 Wind Profile Exponent (p) as a Function of Atmospheric Stability

Stability Class	Rural Exponent	Urban Exponent
A	0.07	0.15
B	0.07	0.15
C	0.10	0.20
D	0.15	0.25
E	0.35	0.30
F	0.55	0.30

The power law equation may be used to adjust wind speeds over a height range from about 10 to 300 m for input to modelling equations introduced later. Adjustments to heights above 300 m should be used with caution. For release heights below 10 m the reference wind speed should be used without adjustment. The reference height is normally assumed to be at 10 m.

The wind direction approximates the direction of transport of the plume. The variability of the direction of transportation over a period of time is a major factor in estimating ground level concentrations averaged over that period, as follows from Figure 3.2.

Wind speed and direction data are available from the Meteorological Office in the UK and from the National Climatic Data Center (NCDC) in the USA. Wind data are often also recorded at industrial sites and at air quality monitoring sites. It is important that the equipment used to record such data be properly designed, sited and maintained to give data that are reasonably representative of the direction and speed of the plume. Maintenance includes taking account of local obstructions such as tree growth and new buildings.

The Atmospheric Boundary Layer

Early dispersion models were designed for simple calculation, although they are commonly applied nowadays using personal computer systems. One major difference of improved models in increasing use is that they take more account of the variations through the atmospheric boundary layer, plume rise and topography.

It is possible to picture in simple terms that the roughness of the Earth's surface caused by trees, buildings and other obstructions interferes with the flow of air as described previously. The closer to the ground, the greater will be this effect of friction, and so the wind speed can be expected to change with altitude (Figure 3.11).

The atmospheric boundary layer, sometimes called the planetary boundary layer or the friction layer, is that region of the atmosphere that experiences surface effects through vertical exchanges of momentum, heat and moisture. Starting with a qualitative description, this boundary layer is divided into three major sub-layers. The layer nearest to the ground could be called the 'roughness layer'. It represents the region in which turbulence is not fully developed. The depth of this layer is designated by z_0, the roughness length, which can be interpreted as the eddy size of the surface. Schemes are available to relate the roughness length to the type of surface, but it can be calculated approximately as:

$$z_0 = \frac{E}{30}$$

where E is the average height of obstacles in the area under investigation. Some typical values of z_0 are shown later in Table 3.10.

Figure 3.11 Surface Roughness Influences Wind Gradient

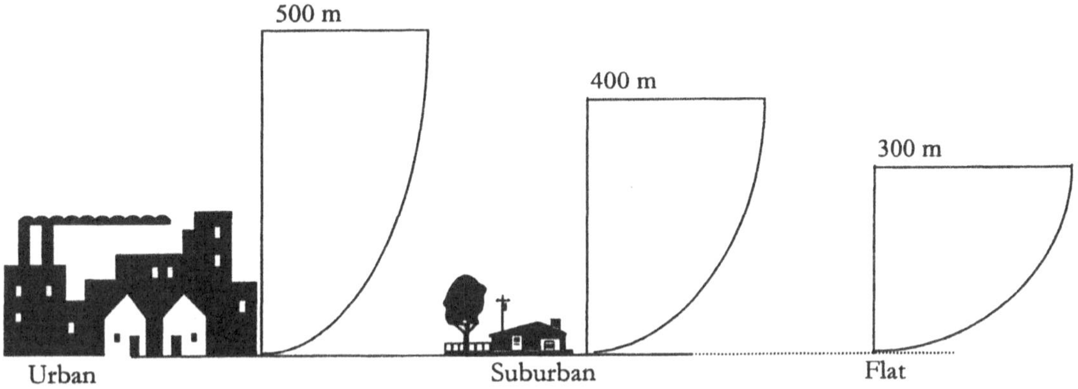

As said at the start of this section, frictional forces at the surface can be viewed as reducing the velocity of the wind at ground level. Above the roughness layer is the surface layer, which extends up to 10–200 m. In this layer, the fluxes of heat, momentum and moisture are assumed independent of height, and the rotational effect of the Earth is negligible. The next layer is the transition layer (also called Ekman layer), extending to between about 100 m and 2 km. The top of this region of the boundary layer, z_i, is the lowest level in the atmosphere at which the ground surface no longer influences the dependent variables in the atmosphere through turbulent mass transfer. Only under special circumstances, such as thunderstorms, does the boundary layer extend to the stratosphere.

The meteorological characteristics of each layer of the atmosphere influence the dispersion of pollutants. In the Ekman layer, turbulent phenomena are the most important. On either side of it, turbulence is not fully developed and molecular motion can play a role in transferring mass and energy. Turbulence is very important because its eddies (ranging from vertical size of 1 mm to the atmosphere boundary layer) can separate parcels of air.

The complex interactions of fluid forces in the atmospheric boundary layer make it difficult to represent accurately the velocity profiles of the wind using simple mathematical expressions. One of the commonly used approaches assumes that the shear stress in the atmosphere is constant irrespective of altitude. In fact, this will only be true for the surface layer up to tens of metres in altitude, but the approach is nevertheless applied to much of the boundary layer. Such a treatment of surface effects leads to another means of considering atmospheric stability. It can be derived that the wind speed at a given altitude is represented by the expression:

$$\bar{u}(z) = \frac{V^*}{k} \ln \frac{z}{z_0}$$

where V* is the frictional velocity with a value of 0.05 to 0.25 m s^{-1}. The von Karman constant, k, is independent of stability and a dimensionless value of 0.4 is experimentally found to be typical. The roughness length, z_0, is calculated from the relationship on page 29. Direct measurement of V* is difficult, and it is usually derived from velocity profile data. The roughness length, as the name implies, is related to the surface roughness. It depends on the size, distribution and geometry of surface elements that range from minute irregularities on ice, to the scale of buildings in a city. This is reflected in the equation for z_0 given previously.

Some values for roughness lengths are summarized in Table 3.11, which also includes selected other parameters for the velocity profile equation.

Table 3.11 Parameters for the Velocity Profile Equation

Surface Configuration	z_0 (cm)	V*/u	u (m s^{-1})	z (m)
Very smooth mud and ice	0.001	0.08	5	2
Deep and even snow cover	0.05			
Bare solid ground	1.0			
Ploughed land	2.0			
Flat open country	3.0	0.071		10
Mown grass (h = 1.4–4.5 cm)	0.2–2.4	0.06–0.09	2–8	2
Long grasses (h = 50–70 cm)	3.7–9.0	0.1–0.32	1.5–6.2	2
Forest	30	0.12		10
Urban area	300	0.22		10

A roughness length varies from 10 to 300 cm across a city like Manchester, but typically has a value of 50 cm. For Milton Keynes a maximum value of about 50 cm has been suggested, with a typical value of 30 cm. A value of 30 cm has been suggested as typical for rural country with small villages, such as may be found in Southern England.* A range of values is given for long grasses in Table 3.11 because grasses bend over easily as the wind speed increases and so a range of values is included for them. Consequently, the value for surface roughness decreases with wind speed.

The first equation presented for surface roughness is applicable only for neutral conditions. In stable conditions, a modified form involves the addition of another term that is linear with altitude. One version of this is:

$$u = \frac{V*}{k}\left(\ln\frac{z}{z_0} + 5.2\frac{(z - z_0)}{L}\right)$$

while for unstable conditions an even more complex equation has been proposed, neither of which need be considered in detail here beyond noting the term L. In the modified equation above, the term L is called the Monin–Obukhov length. This parameter characterizes the 'stability' of the surface layer. It is calculated from widely available measurements and observations at the surface (principally mean wind speed, solar heat flux, cloud cover, roughness length and vegetation type), or from more advanced instrumentation measuring characteristics such as eddy heat flux.

The Monin–Obukhov Length

From the preceding discussion it should be evident that the atmosphere does not flow smoothly, but has apparently random, erratic and rapidly varying motions. This uneven flow is superimposed on the mean flow of the wind. Turbulence is the term used for this highly irregular motion of the wind. Turbulent eddies are produced by two processes:

- thermal turbulence resulting from heating of the atmosphere; and
- mechanical turbulence caused when air moves past an obstruction.

* NRPB-R91 *First Report of a Working Group on Atmospheric Dispersion.* NRPB, Harwell, September 1979

Both occur in any atmospheric situation, but one may predominate. For example, on a clear sunny day with light winds, thermal turbulence will be more important, while mechanical turbulence will dominate on a windy night with neutral atmospheric stability. Simply put, the Monin–Obukhov length is the height above ground at which the production of turbulence by both mechanical and buoyancy forces is equal. Near the surface, shear dominates whereas above the height of the Monin–Obukhov length, buoyancy dominates.

Generally, turbulence is beneficial in terms of enhancing dispersion through the atmosphere, since eddies mix contaminated air with relatively clean air, so reducing pollutant concentrations. However, mechanical turbulence may be a disadvantage where it leads to contaminants being brought close to receptors without significant dilution. This is especially important in the wake of buildings, or when the velocity of discharge from a stack is low compared with the wind speed.

One mathematical definition for the Monin–Obukhov length is:

$$L = -\frac{V^{*3}\, Tc_p \rho}{kgH}$$

where in addition to the terms defined previously, c_p is the specific heat a constant pressure, ρ is the specific gravity, g is the acceleration due to gravity and H is the sensible heat flux between the ground and the air (typically 10 to 120 W m^{-2}). The value of L gives us another measure of the stability of the surface layer of the atmosphere, and can be interpreted into 'stability terms' using Table 3.12.

Table 3.12 Relating the Monin–Obukhov Length to Stability

Monin-Obukhov Length (L)		Stability term
Small negative	$-100 \text{ m} < L < 0$	very unstable
Large negative	$-105 \text{ m} \leq L \leq -100 \text{ m}$	unstable
Very large (+ or -)	$\lvert L \rvert > 105 \text{ m}$	neutral
Large positive	$10 \text{ m} \leq L \leq 105 \text{ m}$	stable
Small positive	$0 < L < 10 \text{ m}$	very stable

Notice that unlike the earlier stability categories, L is a continuous scale. So, a variety of techniques serve to identify atmospheric stability in more objective terms than simply assigning a Pasquill stability category on a qualitative assessment. While these boundary layer issues are not developed further here, they are used in many of the more advanced atmospheric dispersion models currently available. Refined models such as Aermod, ADMS and Indic (Airviro), described in later chapters, process meteorological data to yield values for the Monin–Obukhov length.

Many of the meteorological parameters introduced so far may be summarized in a table in which they are related to stability categories (Table 3.13). Table 3.13 indicates that general patterns of behaviour are evident across the parameters. It also shows that using the traditional Pasquill atmospheric stability category D is most common by far in the UK. This reinforces the point previously made in Table 3.7. It may be noticed that the Pasquill stability varies with surface roughness, which confirms the need to consider not just surface meteorology, but also the conditions at different heights. The outcome of this is that it is incorrect to use the same category for atmospheric stability at different altitudes. From this one may conclude that while a stability category may be assigned for the horizontal plane, the same category may not necessarily apply for the vertical. Moreover, in the horizontal plane the surface roughness may vary. Such factors led to the latest generation of dispersion models such as Aermod, ADMS and Indic (Airviro).

Table 3.13 Meteorological Parameters Related to Stability Categories*

Pasquill stability category	Frequency of occurrence (%)	Mean wind speed (m s^{-1})	Surface heat flux (W m^{-2})	Boundary layer depth (km)	Friction velocity (m s^{-1})	Monin–Obukhov length (m)
A	0.125	0.625	250	1.3	0.11	0.4
A–B	1.25	1.25	180	1.08	0.17	2.5
B	3.8	2.0	150	0.92	0.24	8.4
B–C	2.6	3.37	125	0.5	0.33	26.0
C	15.0	4.12	90	0.84	0.38	55.4
C–D	–	5.0	50	0.6	0.3	115.0
D	62.4	4.12	0	0.5	0.36	∞
E	6.7	3.4	-3.30	0.4	0.26	476
F	–	2.0	-3.33	0.15	0.16	107
F–G	8.4	1.2	-1.39	0.05	0.073	25.6

Terrain Influences

The influence of terrain is an important consideration as the influence on surface roughness has already implied. More than this, the ground level concentration is likely to rise as the land elevation increases. Very steep terrain can cause increased turbulence, whereas deep valleys may promote stagnation conditions. The conclusion has to be that all predictions using models in complex terrain must be treated with caution. Urban influences in terms of their impact on the surface wind speed have already been described, but there are other effects around single buildings as well as across the whole of an urban area that need to be considered.

Building Effects on Dispersion

An air stream approaching an obstruction such as a building is displaced vertically and laterally. A highly turbulent flow is set up and a re-circulation or cavity zone is established in the lee of the building (Figure 3.12). Beyond the cavity zone, a strong mixing motion still exists, although the flow gradually reverts to the undisturbed flow pattern. Beyond this cavity zone is the wake region.

Two aspects are significant in terms of the positive and negative pressure zones and eddies existing around buildings. First, there is the possibility that the release may re-enter the building, and second there is the impact on dispersion. Re-entry is clearly undesirable and discharge above the cavity zone should avoid this problem. As the height of the eddy on the lee of the building extends to 1.3–2 times the building height, a stack should discharge at this height to avoid re-entry. There is still the possibility that the plume may be drawn to ground level in the low pressure region by the effect known as downdraught. One simple rule-of-thumb in use since 1932 suggests that a stack should be 2.5 times the height of any building within a radius of twice the stack height to be reasonably confident that downdraught will not occur, as shown in Figure 3.12. Recent studies suggest that a factor of three is safer. The effect of adjacent buildings will depend on the relative height and location of the stack, and stacks positioned both upwind and downwind of the obstruction can be affected. A more general relationship for the stack height (h_s) necessary to avoid problems on a site that has a simple configuration (for example, a stack adjacent or attached to a single rectangular building), is:

$$h_s = H + 1.5\,L_b\,,$$

* From Hunt J C R et al., 'Developments in Modelling Air Pollution for Regulatory Uses', Proceedings 18th Nato/CCMS, International Conference on Air Pollution Modelling, 1990, Vancouver.

where H is the building height and L_b is the lesser of either building height or maximum projected building width. Tall, narrow buildings and tiered structures present special problems and guidelines are available to deal with them, but are outside the scope of this book.

Further, guidance notes on good practice advise considering the effects of buildings within five times the height of the stack, from that stack. It is possible that the cavity will not extend beyond a site boundary, but a wake may do so. Consequently, the possible impacts of both regions must be considered. This is done in advanced computer versions of the models to be considered later. Many of these models incorporate algorithms to deal with the building cavity.

Figure 3.12 The Influence of a Building on Airflow

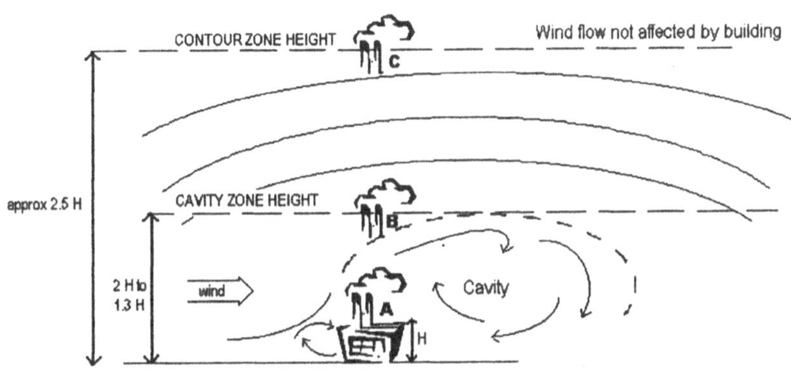

However, a stack is a structure itself, and will therefore interfere with the wind flow. A low pressure region is formed immediately downwind, and this effect, known as downwash, may cause the plume to be drawn into this. Downwash effectively lowers the stack tip and so is to be avoided. Wind tunnel studies have shown that if the efflux velocity is 1.5 times the wind speed, downwash is minimized. Notice from Figure 3.12 that an effect on wind speed is implied, which relates to the concept of surface roughness discussed previously. The drag of the ground reduces the wind speed and so the taller the stack the greater the windspeed at its top.

Other obstacles to airflow will have an influence on dispersion. For example, fences will be important when considering the dispersion of ground level releases, such as may occur following the rupture of a storage tank. It is difficult to envisage such obstacles resulting in ground level concentrations greater than would occur in the absence of the obstacle when considering releases of dense gases. Evidence suggests that the upwind concentrations increase, while those beyond a fence are significantly reduced. However, it is important to consider each situation separately, since there may be special circumstances leading to unusual behaviour.

The Urban Heat Island

Any fuel whether it is coal, oil, gas, electricity or other source of energy is ultimately degraded to heat which is lost. The sensible heat flux from an urban area causes an increase in the temperature of an air mass as it passes over the urban–rural interface. A problem in assessing the effect of the urban heat island, as this effect is known is to estimate the net sensible heat flux of the urban zone over its rural surroundings. Figure 3.13 illustrates this effect, drawing on measurements made in a section across Melbourne.

Many factors contribute to this effect, including the balance between solar heating and heat from human activity, as well as the heat capacity and albedo (% reflectivity) of the surface. One outcome taking account of the effect assumes that during the night in moderate winds, a downward flux (negative) of about 2.5 mW cm^{-2} will cause a stable or very stable stratification of the atmosphere. A shift towards 1.5 mW cm^{-2} is thought to be sufficient to move the atmospheric stability one category towards neutral. Thus, in terms of the Pasquill stability categories, F becomes E and E becomes D.

So, in order to maintain an urban heat island effect sufficiently great to change the atmosphere over a city to stable conditions, the city must maintain an upward heat flux of about 1mW cm^{-2} compared with its surroundings. Evidence suggests that this is indeed the case for urban areas. Hence, in an urban setting it is necessary to move stable categories one step towards neutral stability. The heat island effect is much less significant in neutral and particularly unstable conditions, and so may be neglected in such cases. Urban effects are recognized in dispersion models, which require the user to identify the topography for the scenario being modelled. Systematic ways of doing this will be described later.

Figure 3.13 Urban Heat Island Effect: Temperatures in a Section Across a City

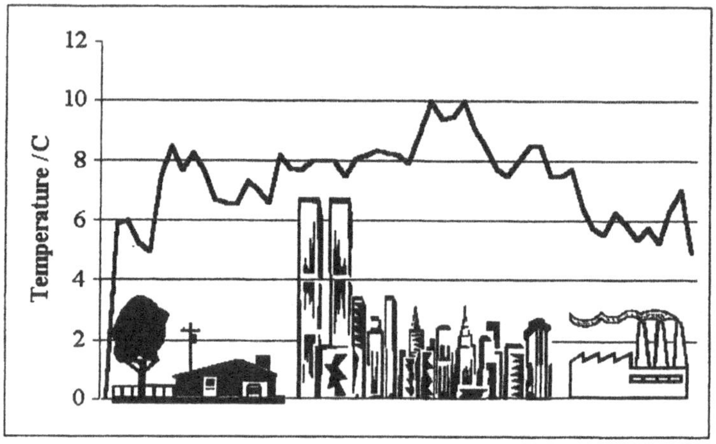

The Sea Breeze

The sea breeze can be an important mesoscale weather phenomenon along coastlines around the world, including the shoreline of major lakes. It is a thermally driven circulation that occurs as a result of the differences in heat capacities of water and the land. The land more readily absorbs and radiates solar radiation than water surfaces. When the sun rises and solar radiation heats the land, air masses over the land become more buoyant and they rise as they are warmed by the radiation. The air over the surface of the water is cooler during the daytime since the heat capacity of water is greater than that of the land. The vertical pressure gradient is greater in this cool air than over the warmer land mass due to hydrostatic conditions. This pressure gradient produces a light wind flow from the sea to the land as the buoyant air parcels move towards the lower pressures over the land. This is the 'sea breeze' that is felt at the surface especially during the afternoon, as the winds blow onshore from the cooler water. When the sun sets, the opposite circulation develops as the land radiates energy more quickly than the sea, and at the surface, the wind blows from the cooler land mass towards the relatively warmer sea surface. These 'land breezes' are weaker and tend to lead to lesser convergence over the sea than the sea breeze induces over the land. This is primarily because solar radiation is not present to add the potential energy needed to increase lifting over the sea surface during the night.

The onset of a sea breeze is marked by changes in wind speed, temperature and humidity at a specific location as the sea breeze 'front' passes through. At the leading edge of the circulation, the wind shows an abrupt shift direction onshore. There is an increase in wind speed and humidity, and a decrease in temperature as the air parcels moistened above the sea surface blown onto the land.

The vertical depth of the sea breeze is generally below three kilometres. The stability of the lowest layers of the troposphere decreases as the air parcels move onshore and are lifted by the warmer land mass. The circulation pattern that sets up extends as far as 20 km inland when the onshore winds begin, usually around noon when the land becomes warmer than the sea surface.

There are several other factors that effect the sea breeze circulation when it develops. The Coriolis force that is present due to the Earth's rotation helps to turn the winds throughout the duration of the day. A sea breeze that begins as a direct onshore wind at the immediate coast is often turned parallel to the shore by evening, when the thermal gradients diminish along with the winds. As with most mesoscale systems, the sea breeze winds depart a great deal from the geostrophic wind. In the vertical, the Coriolis force acts to veer the sea breeze winds with height. Topography can also influence the sea breeze. Hills and valleys near the coast can increase the velocities associated with the circulation and change the wind direction. A coastline with little vegetation cover will allow the land surface to heat faster, making more energy available for the onset of the sea breeze

The sea breeze can also affect pollution in cities located near coasts. The sea breeze often pushes polluted air away from urban centres and can concentrate pollution in areas further inland. Modelling air quality in such locations requires special considerations with modified algorithms.

Conclusion

Air quality changes continuously and concentrations tend to follow a log-normal distribution. This needs to be recognized in quoting representative values and comparing data against each other or against standards, as does the importance of averaging time on concentration.

Studies of transport and dispersion near the surface necessarily involve consideration of surface and atmospheric heterogeneity. Much work has been done to improve the understanding (and depictions) of the factors that influence the planetary boundary layer, especially the central considerations of heat energy and topography. For example, local breezes, such as urban heat island or sea-breeze/land-breeze circuits, may partially mask the wind from the circulation around the migratory highs and lows. Both the sea-breeze/land-breeze circuit and the valley-breeze/mountain-breeze system are a result of differential local heating. Increasingly, atmospheric models will combine weather forecasting with the effects of local factors that influence the planetary boundary layer. Air quality models have demanded this kind of coupling for a considerable time and the latest generation of models use well-developed descriptions of boundary layer processes.

Exercises

1. a) If the mean daily mean concentration of an air pollutant at a given site is 0.067 ppm and the SGD is 1.88, what is the geometric mean for 1 hr?
   b) What is the yearly mean when the daily mean is 0.046 ppm with a SGD of 1.72?

2. Predict the atmospheric stability category based on the following observations.
   a) On 28 December, 1 hr of total bright sunshine and the mean wind speed of 5.1 km h^{-1}.
   b) On 23 August, there were 11.4 hr of bright sunshine and a mean wind speed of 11.9 km h^{-1}.
   b) On 27 September, 3 hr of bright sunshine and a mean wind speed of 28.8 km h^{-1}.

3. In a study involving monitoring and modelling of nitrogen oxides in the 'East Thames Corridor' maximal 1 hr concentrations were derived from hourly mean concentrations. Data from continuous chemiluminescent analysers in Central London and West London indicated a range of standard geometric deviations of 1.3 to 1.6. What factors would be used to estimate 1 hr maximum concentrations from the annual mean values?

4. The following data represent the environmental lapse rate observed by a radiosonde ascent. Compare the lapse rate with the dry adiabatic lapse rate and comment on the implications for dispersion.

Altitude (m)	Environmental Temperature (°C)	Altitude (m)	Environmental Temperature (°C)
0	12	800	6
100	10.8	900	6
200	9.6	1000	6
300	8.4	1100	6.976
400	7.2	1200	7.952
500	6	1300	8.928
600	6	1400	9.904
700	6	1500	10.88

5. The following data represent mean temperatures recorded on 14 ascents at midday in Oman, and observations from a radiosonde ascent in East Sussex at midnight on 8 January 1998. Comment on the implications of these conditions for the dispersion of pollution.

Oman		Sussex	
Altitude (m)	Temp (°C)	Altitude (m)	Temp (°C)
0	27.5	0	5.5
400	25	100	7
1100	21	200	5.5
1600	23	400	5.5
2200	24	1100	2.5
2800	20	1600	0
3500	15	2200	-2
4000	10	2800	-7.5
4500	5	3400	-11
4700	-5	3500	-11
		3600	-11
		4000	-13
		4500	-18
		4700	-28

CHAPTER 4
EMISSION INVENTORIES

It is evident from the discussion early in Chapter 2 that a primary input to a model is data on emissions from sources under investigation. Hence, it is appropriate to consider how such information is obtained.

Source Data

There are many possible sources of emissions to the atmosphere from so-called stationary sources. These may be categorized as planned releases and fugitive releases. The former come from stacks, vents, ducts, flares, storage tanks, cooling towers etc. Fugitive releases include those from valve glands, pump seals, flanges, compressor seals, sampling points, wind losses from stock piles, as well as plant losses from welding, molten metal handling etc.

In order to estimate the impact of a stationary point or area source on air quality, certain characteristics of the source must be known. In general, the following information is needed:

- Pollutant emission rate.
- Location.
- Stack height for a point source and release height for an area source.
- Dimensions of an area source or a line source.
- Stack gas temperature, exit velocity and stack inside diameter (for plume rise calculations).
- The character of the surrounding topography, and the location of the emission point within it.
- A description of all structures in the vicinity of (or attached to) the stack in question.
- Similar information from other significant sources near the subject source.

For nearby sources, air quality data or dispersion modelling results that demonstrate the air quality impact of those sources will suffice to provide background data. The analysis of air quality impact requires that the emissions from each source be fully and accurately characterized. Temporal variations in emissions should be considered. Failure to do so leads to errors in the predictions, as the examples in later chapters will illustrate. Options for acquiring emission data are as follows:

- Indirect determination: mass balance or rigorous calculation; design or manufacturers' test data.
- Emission factors and activity data.
- Engineering judgement.
- Reference to permitted levels in authorizations as recorded in public registers etc.

If pollutants are not emitted at a constant rate, and most are not, information should be obtained on how emissions vary with season, day of the week and hour of the day. In most cases, emission rates vary with the source production rate or rate of fuel consumption. For example, for a coal-fired power plant selling electricity to the national grid, emissions are related to the kilowatt-hours of electricity produced, which is proportional to the amount of fuel used to produce the electricity. At a minimum, impact estimates should be made with source characteristics representative of the design capacity (100% load may be rare) and operating hours. The impacts should be estimated based on source characteristics at loads of 50% and 75% of design capacity, and the maximum impacts selected for comparison with the relevant air quality standard or other relevant criterion.

In addition, where emissions are reduced by abatement equipment, the effectiveness of abatement must be accounted for in the emissions analysis. The operators should be able to estimate the effectiveness of controls in reducing emissions and in altering emission characteristics such as efflux temperature and velocity. They should also know how the effectiveness varies with changes in plant operating conditions.

Within an organization it may be necessary to improve the monitoring capability in order to establish the required data. For fugitive releases that escape collection by hoods, there may be no data. In some situations, the cost and inherent uncertainties associated with monitoring may suggest alternative approaches, such as deriving a mass balance. This also serves as a means for cross-checking release monitoring. Attempting to produce a mass balance will reveal deficiencies, and will indicate where available information has been based on calculations rather than on direct emissions. The ultimate aim is to derive an inventory of releases and their sources. This inventory includes:

- release points;
- substances;
- physical measures – rate, concentration, mass, temperature, particle size etc;
- measurement methods;
- time variability; and
- statistical parameters.

Some guidance on monitoring is available*, but the final decision on what technique is required depends on the regulator and the specific process, its age and abatement equipment within the philosophy of the best available techniques (BAT).

It is also possible to use emission factors. These serve an important role in deriving emission inventories for modelling. Fugitive emissions from an area source are likely to vary with wind speed and both atmospheric and ground moisture content, especially for open ground or mineral workings. If pollutant emission data are not directly available, emissions can be estimated from fuel consumption or production rates by multiplying the rates by appropriate emission factors. Emission factors can be determined using three different methods. In decreasing order of confidence they are:

1. Stack test data or other emission measurements from an identical or similar source.
2. Material balance calculations based on engineering knowledge of the process.
3. Emission factors derived for similar sources or obtained from a compilation.

It is impracticable to measure every emission source in a large urban area. The majority of emissions are therefore estimated from relevant local information such as fuel consumption, vehicle kilometres travelled, or other measure of activity relating to the emissions. Emission factors, derived from the results of measurements, are then applied to the activity data in order to estimate the likely emissions:

$$\text{Activity rate} \times \text{Emission factor} = \text{Emission rate}$$

Emission factors used in preparing urban emission inventories are derived from various sources:

- Emission factors developed in the UK based on measurements made at industrial and other plants as well as of vehicles and equipment.†

---

* a) HMIP Technical Guidance Note M1 *Sampling Facility Requirements for the Monitoring of Particulates in Gaseous Releases to Atmosphere*, HMSO, 1993.

b) HMIP Technical Guidance Note M2, *Monitoring of Emissions at Source*, HMSO 1993.

† The UK Emission Factors Database, http://www.rsk.co.uk/ukefd/index.htm

- The European Environment Agency's (1996) Atmospheric Emissions Inventory Guidebook* and more recent European studies.
- The US Environmental Protection Agency's (1995) manual on the Compilation of Air Pollution Emission Factors, generally known as AP-42.†

Where possible, emission factors derived from measurements undertaken within the country itself are preferred. In cases where factors derived from national data are not available, factors developed elsewhere are used, but there may be some doubt as to their appropriateness. While, as noted earlier, the number of emission sources that have been continuously measured is limited, in many cases enough sources have been studied to derive robust emission factors. Measurement of further emission sources and additional characteristics of emissions are constantly adding to the stock of available data. Some examples of other sources of information on fugitive releases are as follows:

- *Protocol for Equipment Leak Emission Estimates*, EPA-453/R-93-026.
- Reference Method 21, *Determination of Volatile Organic Compound Leaks*, US EPA Title 40, Part 60, Appendix A, 1990.
- *Improving Air Quality, Guidance for Estimating Fugitive Emissions from Equipment*, Chemical Manufacturers Association, Washington, January 1989.
- *Fugitive Hydrocarbon Emissions from Oil & Gas Production Operations*, API 4589.
- *Evaporative Losses from Storage Tanks*, API 2517, 2518, 2519.
- *Emission Estimation Technique Manual for Fugitive Emissions*, National Pollutant Inventory, Environment Australia, December 1999.

An emissions inventory is a schedule of the sources of an air pollutant or pollutants within a particular geographical area. The inventory usually includes information on the amount of the pollutant released from major industrial sources, average figures for the emissions from smaller sources, and from transport throughout the area. Emission inventories are an essential tool in the management of local air quality. While monitoring such as that carried out through the UK Automatic Monitoring Networks gives data on ambient air quality, emission inventories identify the sources of pollution and help in preparing abatement strategies.

By identifying the sources of air pollutants, an atmospheric emissions inventory can be used as an aid in interpreting air quality measurements. As pointed out in Chapter 1, while there are a number of very well established techniques for measuring the concentrations of pollutants in the atmosphere, it is not always possible to identify their sources or the relative importance of multiple sources. The use of an emissions inventory may be the only way of doing this. Knowledge of the significant sources of air pollutants is essential before any action can be taken to improve air quality in a particular area under the legal framework provided by Part IV of the Environment Act 1995.

The starting point for local authorities is the review of local air quality and its assessment against standards and objectives. The results of these reviews by local authorities will determine what, if any, further action may be required, such as the establishment of an Air Quality Management Area. Quality assessments include the use of three main tools: air quality monitoring, local emission inventories, and numerical modelling.

In the UK, as part of the government's continuing programme of air pollution studies, it has developed a national inventory of air pollution sources and the type and quantity of the pollutants they emit. This is called the National Atmospheric Emissions Inventory (NAEI). The first NAEI report was prepared in 1987 but emission estimates have since been prepared retrospectively to 1970

* *Atmospheric Emissions Inventory Guidebook*. First Edition. Produced Jointly by EMEP/CORINAIR. Prepared by the EMEP Task Force on Emission Inventories. Edited by Gordon McInnes. European Environment Agency, February 1996.

† *Compilation of Air Pollutant Emission Factors* (AP-42). Published by the Office of Air Quality Planning And Standards, Office Of Air And Radiation, US Environmental Protection Agency (US EPA). November 1996.

as well as for recent years. The NAEI estimates the annual emissions resulting from both fuel combustion and from non-combustion sources, and these are categorized in accordance with the UNECE/CORINAIR SNAP (Selected Nomenclature for Air Pollution) classification and by source type. The coverage of this inventory has been expanded as emissions of pollutants have grown or as evidence of their adverse effects has accumulated. Initially only emissions of black smoke and sulphur dioxide (SO_2) were estimated but now ten pollutants or pollutant groups are covered. They are carbon dioxide (CO_2), SO_2, black smoke, carbon monoxide (CO), oxides of nitrogen (NOx), methane (CH_4), non-methane volatile organic compounds (NMVOC), ammonia (NH_3), nitrous oxide (N_2O), lead, other heavy metals and halogens. In addition, HFCs and PM_{10} are being considered for inclusion in the inventory. The National Environmental Technology Centre maintains the NAEI for the government and data are available through the Internet (Appendix 5).

The purpose of the NAEI is to provide:

- input to discussions with various international bodies;
- input to UK policy-making with respect to pollution abatement and control;
- assistance in judging the effectiveness of existing policies;
- an aid to the interpretation of air quality measurements;
- input to atmospheric dispersion models; and
- general public information.

Data are collected for three types of sources:

1. Line sources including roads and railways.
2. Area sources including emissions from agricultural and other land, and low intensity emissions from sources such as building heating systems.
3. Point sources including emissions from industrial plants.

The NAEI has considerable potential to assist local authorities in the reviews of local air quality that they are required to carry out under the Environment Act 1995. To be of greatest value to local authorities, since 1998 the NAEI has included criteria pollutants at the resolution of a 1 × 1 km grid.

Conclusion

Source data are an essential input to any air quality model, and ideally should reflect the variations in emissions with time. Much data are available in public registers and in data-bases of emissions and emission inventories. Increasingly these will be developed to provide high quality input to models. Box models can serve to validate inventories. Ultimately, the reliability of any model output depends on the quality of data input and by improving the quality of emission data, one source of uncertainty in modelling can be reduced.

CHAPTER 5
GAUSSIAN PLUME DISPERSION MODELS

Material discharged into the atmosphere is transported by the wind and mixed into the surrounding air by the turbulent eddies introduced earlier, and this process causes dilution of the pollutants. The extent of this dilution is a function of the following characteristics:

- the physical nature of the pollutant, ie solid or gas;
- the physical characteristics of the atmosphere; and
- the physical characteristics (temperature, density, velocity and height) of the emitted effluent.

These features are interactive and the level of pollution observed at a location some distance away from the source will be a function of all three. However, with knowledge of the physical processes involved, it is possible to develop mathematical equations capable of predicting these levels, ie to model the dispersion. Because of the complexity of the interactions, it is convenient to start by making some simplifying assumptions. The theory can then be progressively extended to cover the additional complications present in any realistic situation.

When ground level concentrations of pollutants from known sources can be predicted with sufficient accuracy, the knowledge can be applied to improve the environment. The consequences of alterations in the nature of emissions from existing sources or the introduction of new sources can be assessed and the relevant authorities can specify appropriate conditions for discharge into the atmosphere.

The most commonly used air pollution models are probably Gaussian plume dispersion models that predict ground level concentrations of conservative pollutants within a few kilometres of the source. Many texts have dealt with such models in greater or lesser detail and there is no intention of repeating much of that here. The aim of this chapter is simply to outline this type of model and, for completeness, to provide sufficient knowledge to understand their use in practical applications.

A Frame of Reference

The definition of the three dimensional system must be considered as a fundamental starting point. The air around us and the components within it can move in any direction, so how is it possible to model the impact on air quality at a given point in space? The horizontal distance moving in the wind direction away from the source defines that point in space (Figure 5.1).

The distance from the source is unbounded in the positive. That is, the pollutant can move as far in the positive x direction as possible. Thus, x is usually a positive number with the source being the zero point. Bear in mind that eddies and turbulence may make x negative. The lateral distance of dispersion from the centre line of the wind direction is denoted by y, which is also unbounded in either direction. So y is measured as a positive or a negative distance from the centre line which is always along the line of the wind direction.

Figure 5.1 The Dimensions for Dispersion

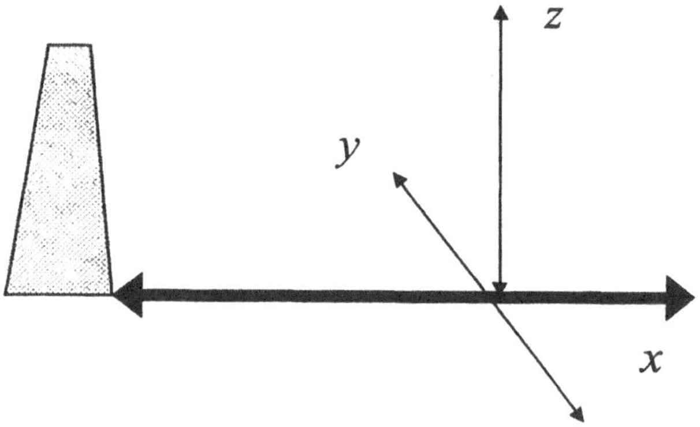

The ground sets a boundary on the vertical dispersion as the pollutant moves away from the source. This boundary means that once the pollution reaches the ground it may be absorbed as fallout and leave the system. Thus z is always measured as the vertical distance from the ground, but this is complicated somewhat by the fact that the release may occur above the ground level. So, z may have positive or negative values from the release height. This release height may be the stack height. This will be discussed in more detail later on.

Thus, the goal of a modelling exercise would be to determine the concentration of pollution at any given point denoted by the coordinates (x,y,z). To do so a number of parameters are needed. Primarily these are source characteristics and how the plume spreads as it travels through the air.

An alternative reference system follows the motion of the atmosphere. The two options are represented in Figure 5.2.

Figure 5.2 Reference Systems

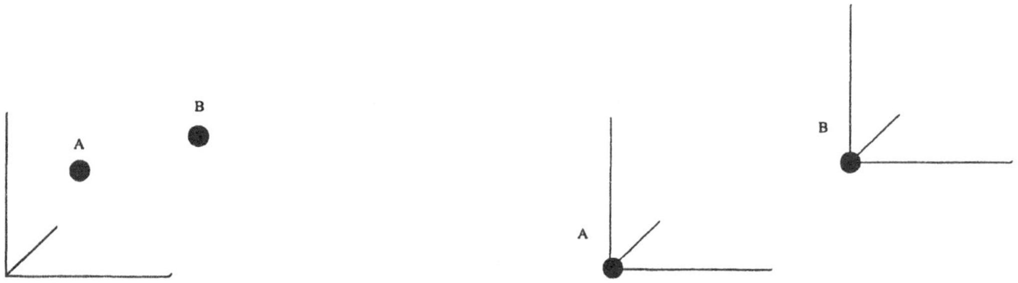

(a) fixed **(b) moving**

Eulerian and Lagrangian models both solve the same advection-diffusion equation. In both cases, a hypothetical parcel of air is considered moving from position A to position B. In Figure 5.2(a), the reference system is fixed with respect to the earth and a grid is laid out over the region of interest. This Eulerian reference system is used for most of the simpler models. The solution is obtained relative to a fixed reference point or grid, against which both pollutant concentrations and meteorological variables are defined:

Eulerian Dispersion Model: dC/dt = [advection] + [diffusion] + [source] + [sinks]

The reference system represented in Figure 5.2(b) is the basis of Lagrangian systems, in which the coordinate system follows the flow of the air. In fact, particles have a Lagrangian nature since they move to follow the main flow, and an especially powerful computational tool has been the development of dispersion models based on hypothetical 'particles' to simulate the dynamics of atmospheric dispersion. In the Lagrangian frame, the advection diffusion equation is rewritten such that the change of concentration is determined following the parcel as it is transported by the wind.

Lagrangian Dispersion Model: DC/Dt = [diffusion] + [source] + [sinks]

The two methods yield identical solutions. While both approaches have been successfully applied to a variety of different simulation configurations, Eulerian approaches are more suitable when complex emission and non-linear chemical conversions are involved. Lagrangian models are most applicable to point-source simulations. In a Lagrangian particle model, releasing many particles over the duration of the release simulates the source. In addition to the mean advective motion of each particle (ie its trajectory), a random component to the motion is added at each step according to the atmospheric turbulence at that time. In this way a cluster of particles released at the same point will expand in space and time simulating the dispersive nature of the atmosphere. Air concentrations are calculated by summing the mass of all the particles in a grid cell.

Several of the latest generation of dispersion models use the Lagrangian system, although they are not strictly Gaussian plume models. As the foregoing description suggests, Lagrangian models are more complex than those based on Eulerian reference systems, and examples are outlined in more detail later. In addition, advanced models consider the atmosphere using a physical approach rather than the subjective, empirical approach based on Pasquill stability categories. Such differences were introduced in Chapter 3.

Gaussian Plume Models

In Gaussian plume models, the cross-wind plume concentration distributions are taken to be Gaussian in form. This has been partially substantiated through field experiments for typical meteorological conditions.

Imagine a single point source of release to the atmosphere, such as may occur from a single stack. After emission, the pollutants mix with the air and the plume begins to expand steadily in size. The next phase of the dispersion process depends entirely on the meteorological characteristics such as those introduced previously in Chapter 3. Observing a plume emanating from the stack at any instant in time would indicate that its shape is irregular. Moments later, the shape may be very different, but over an extended period the plume can be envisaged to be contained within an overall envelope represented diagrammatically in Figure 5.3. The implications for concentration as a function of averaging time were discussed in Chapter 3.

Figure 5.3 Generalized Plume Envelope

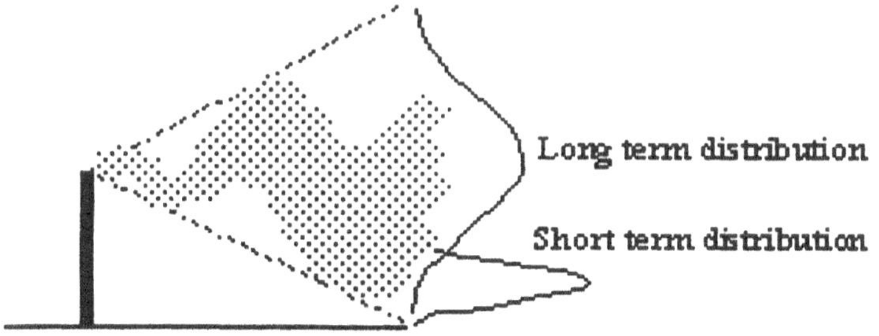

A concentration profile is drawn for a representative distance away from source. The profile represents the following detail.

- Plumes are three-dimensional. Pollutants are dispersed vertically and horizontally at 90° to the wind direction. The amount of horizontal dispersion is normally slightly greater than the amount of vertical dispersion because the latter is limited by the ground and the top of the mixing layer.
- Plumes increase in cross-section with increasing distance away from source. This is reflected in a broadening concentration profile. Both statements are really ways of saying that pollutants are diluted as they move away from the source.

The shape of the concentration profile shown in Figure 5.3 conforms closely to the Gaussian or normal distribution curve. In the strict sense, Gaussian diffusion is valid only for long diffusion time and for homogeneous, stationary conditions. However, this type of model has been found to give useful results for many practical applications. The Gaussian plume algorithm has advantages of inherent simplicity, ease of use, flexibility and short computation time. But, it also has some disadvantages:

- Concentrations are not time-dependent in the usual sense (the approach is 'quasi-steady-state' in that the input variables are normally updated only once per hour).
- Spatial variability in the meteorological parameters is difficult to incorporate.
- Difficulties are encountered when the wind speed is light and wind direction is poorly defined.
- The simple approach cannot be used for reactive or secondary pollutants, because in these cases superposition of individual source contributions at a receptor is not valid.

Two variants of Gaussian models may be considered. The first is known as the plume model, and describes the steady-state concentration of material released from a continuous source such as a stack. The second type is the puff model. This describes the concentration of material from a single release of a fixed amount of material. The distinction between the two types of release is that a steady state plume is generally formed downwind of a release from a stack, whereas a puff model scenario may be a sudden release of a finite amount of material from a rupture in a storage tank. This produces a vapour cloud that moves away from the source. In reality a plume may be regarded as a continuous series of puffs, but a plume model is easier to use for describing plume behaviour. The puff model is useful in dynamic situations, such as when there is a change of wind direction.

A starting point is the instantaneous release of a fixed mass of material, Q, into an unbounded atmosphere. The coordinate system is conventionally based at the source. As shown in Figure 5.1, the x-axis is the centre line directly downwind from the release point, and rotates according to the wind direction. The y-axis is the distance from the centre line and the z-axis is the elevation from ground zero or the release point as specified.

If it is assumed that no chemical reactions occur and there is no molecular diffusion, the concentration of material, C, resulting from the release is given by the advection equation

$$\frac{\partial C}{\partial t} + \frac{\partial}{\partial x_i}\left(u_i C\right) = 0$$

In this equation, u_i is the air velocity and the subscript i represents the sum over all coordinate directions x, y and z. If the velocity in this equation is made equal to the average wind velocity and the equation is solved, it is found that dispersion occurs much faster than predicted. This is due to turbulence in the velocity field. If it were possible to specify exactly the wind velocity according to time and position, the equation would be more valid, but turbulence is difficult to include in simple models, and so simplifications are often used. If a mean value $\langle u_i \rangle$ and a stochastic value u_i' are used to represent the velocity, then:

$$u_i = \langle u_i \rangle + u_i'$$

Therefore, the concentration will also fluctuate as a result of this velocity field, so:

$$C = \langle C \rangle + C'$$

In this case ⟨C⟩ is the mean concentration and C' is the stochastic fluctuation. As the fluctuations in both C and u_I are around the mean values, it follows that $\langle u_i \rangle = 0$ and $\langle C' \rangle = 0$. Substituting the equations for u_I and C into the advection equation and averaging the result over time gives:

$$\frac{\partial C}{\partial t} + \frac{\partial}{\partial x_i}\left(\langle u_i \rangle \langle C \rangle\right) + \frac{\partial}{\partial x_i}\left(\langle u_i' \rangle \langle C' \rangle\right) = 0$$

When averaged, the terms $\langle u_i \rangle C'$ and $u_I' \langle C \rangle$ become zero, but the turbulent flux term $\langle u_I' C' \rangle$ is not necessarily zero, and so remains in the equation. An additional equation is needed to describe the turbulent flux, and the usual approach for this is to define an eddy diffusivity, K_i such that

$$\langle u_i' C' \rangle = -K_i \frac{\partial \langle C \rangle}{\partial x_i}$$

Combining the equations yields:

$$\frac{\partial C}{\partial t} + \frac{\partial}{\partial x_i}\left(\langle u_i \rangle \langle C \rangle\right) = \frac{\partial}{\partial x_i}\left(K_i \frac{\partial \langle C \rangle}{\partial x_i}\right)$$

If it is assumed that the atmosphere is not compressible, then $\dfrac{\partial \langle u_i \rangle}{\partial x_i} = 0$ and so

$$\frac{\partial \langle C \rangle}{\partial t} + \langle u_i \rangle \frac{\partial \langle C \rangle}{\partial x_i} = \frac{\partial}{\partial x_i}\left(K_i \frac{\partial \langle C \rangle}{\partial x_i}\right)$$

When combined with appropriate initial conditions and boundary conditions, this equation forms the basis for dispersion modelling, and may be solved for various puff and plume scenarios. Similar equations can represent the three-dimensional situation of real life. In this equation transportation due to the wind and diffusion due to turbulence are both represented.

In order to solve this equation, it is necessary to make several assumptions:

- the rate of emissions from the source is constant, and so the solution is not variable with time;
- the wind speed is constant over time, position and elevation;
- the position does not influence the diffusivity;
- diffusion along the *x* coordinate is insignificant compared with the flow along that direction;
- the pollutant is conserved – thus, there is no loss by decay, chemical reaction or absorption; and
- when the pollutant reaches the ground it is not absorbed but is reflected.

In general terms, the eddy diffusivity varies with weather conditions, location and time, and so is less useful in a practical sense. Sutton* resolved this difficulty by proposing the concept of a dispersion coefficient. The three dispersion coefficients σ_x, σ_y, and σ_z represent the standard deviations of the concentration in the downwind (x), cross-wind (y) and vertical (z) planes respectively. Values for dispersion coefficients are more easily obtained than values for eddy diffusivities. The larger the value of σ, the broader the distribution. Consequently, further from the source the plume becomes broader and so the value of σ is greater. Putting this another way, the value of σ depends on the downwind distance x. Ways of determining their values will be described shortly.

* Sutton O G, *Micrometeorology*, McGraw Hill, New York, 1953, p286.

An instantaneous point source is a purely theoretical scenario, but is considered as an example of a puff release. For a constant wind direction and velocity, u, the solution for the Gaussian dispersion equation is:

$$\langle C\rangle(x, y, z, t) = \frac{Q}{\sqrt{2}\pi^{3/2}\sigma_x\sigma_y\sigma_z}\exp\left\{-\frac{1}{2}\left[\left(\frac{x-ut}{\sigma_x}\right)^2+\frac{y^2}{\sigma_y{}^2}+\frac{z^2}{\sigma_z{}^2}\right]\right\}$$

A continuous release may be considered as an infinite number of overlapping puffs carried along in the wind at velocity u. One basic assumption is that diffusion along the *x*-axis is insignificant compared with gross transport in the wind. In addition, as Figure 5.3 implies, when a plume disperses from an elevated source, it spreads vertically until eventually the lower boundary reaches the ground. Further downward diffusion is impossible, but it is assumed that there is no loss at the surface. The real-life situation is well represented by the assumption that the plume is reflected. Effectively this may be viewed as a virtual source at an effective height, H, below the ground. This mathematical contrivance to allow for reflection leads to the concentration of a pollutant in the air therefore being represented by:

$$C\,(x, y, z) = \frac{Q}{2\pi u\sigma_y\sigma_z}\exp\left(-\frac{y^2}{2\sigma_y{}^2}\right)\left(\exp\left(-\frac{(z-H)^2}{2\sigma_z{}^2}\right)+\exp\left(-\frac{(z+H)^2}{2\sigma_z{}^2}\right)\right)$$

The net effect of introducing multiple reflections is that the airborne concentration at any given point will be the summation of the contributions from the true and several virtual sources. The number of contributions will depend on the vertical spread of the plume as well as on the height of the mixing layer, since an inversion may also cause reflection. In practice, it has been found that including the contribution from one virtual source below ground and, if necessary, one at the top of the boundary layer or mixing height (M), gives sufficient accuracy. The dispersion equation becomes:

$$C=\frac{Q}{2\pi u\sigma_y\sigma_z}\exp\left(-\frac{y^2}{2\sigma_y^2}\right)\left[\exp\left(-\frac{(z-H)^2}{2\sigma_z^2}\right)+\exp\left(-\frac{(z+H)^2}{2\sigma_z^2}\right)+\exp\left(-\frac{(2M+z+H)^2}{2\sigma_z^2}\right)+\exp\left(-\frac{(2M+z-H)^2}{2\sigma_z^2}\right)+\exp\left(-\frac{(2M-z+H)^2}{2\sigma_z^2}\right)+\exp\left(-\frac{(2M-z-H)^2}{2\sigma_z^2}\right)\right]$$

In practice, it is not necessary to use this extended equation, although it is important to recognize the significance of an inversion layer on air quality and the mathematical model.

This equation describes the concentration everywhere in the atmosphere. However, for routine use a number of simplifications are possible for the most common situations. Generally, the concern will be over concentrations at ground level, whether for considerations of public health, injury to vegetation or damage to materials. At ground level, z = 0. Substituting z = 0 gives:

$$C=\frac{Q}{\pi u\sigma_y\sigma_z}\exp\left(-\frac{y^2}{2\sigma_y^2}\right)\left[\exp\left(-\frac{(H)^2}{2\sigma_z^2}\right)\right]$$

Sometimes this is expressed in the equivalent form:

$$C=\frac{Q}{\pi u\sigma_y\sigma_z}\exp-\left(\frac{y^2}{2\sigma_y^2}+\frac{(H)^2}{2\sigma_z^2}\right)$$

Any consistent set of units may be used. Models such as this may also be used for an approximate representation of the particle dispersion for particles below about 30 μm in diameter, since they behave in a similar manner to gas molecules. Larger solids are subject to deposition forces, and terms to account for these can be incorporated in more advanced models.

Often the concern will be over the maximum concentration at ground level and this is most likely to occur directly under the plume centre line. So, a further simplified equation gives the maximum concentration at ground level under the plume centre line. At ground level (z = 0), and under the plume centre line where y = 0, substituting these values gives:

$$C=\frac{Q}{\pi u\sigma_y\sigma_z}\left[\exp\left(-\frac{(H)^2}{2\sigma_z^2}\right)\right]$$

For a source at ground level, such as a bonfire or for very fine dust from a stockpile, H = 0. So:

$$C = \frac{Q}{\pi u \sigma_y \sigma_z}$$

Where the concern is to find the maximum downwind concentration and its location, a procedure for this is to use the following equation to indicate the maximum downwind distance:

$$\sigma_z = \frac{H}{\sqrt{2}}$$

Then the maximum concentration at ground level along the x-axis, $\langle C \rangle_{max}$ is given by:

$$\langle C \rangle_{max} = \frac{2Q}{e\pi u H^2} \exp\left(\frac{\sigma_z}{\sigma_y}\right)$$

Formulae have been derived for line sources based on similar assumptions to those for point sources. However, differences include the treatment of vehicle wakes and the dispersion coefficients. By way of example, the concentration at a receptor located at (x, y, z) resulting from a line source at height h and with a continuous emission rate of Q g m^{-1} s^{-1} and parallel to the y-axis is given by:

$$C = \frac{Q}{\sqrt{2\pi}\,\sigma_z u}\left\{\exp\left[-\frac{(z-h)^2}{2\sigma_z^2}\right] + \exp\left[-\frac{(z+h)^2}{2\sigma_z^2}\right]\right\}$$

This assumes that the wind direction is perpendicular to the road and the source has an infinite extension along the y-axis. Hence, the concentration does not depend on the value of y.
When the source is at ground level, h = 0 and so:

$$C = \frac{2Q}{\sqrt{2\pi}\,\sigma_z u}\left\{\exp\left[-\frac{z^2}{2\sigma_z^2}\right]\right\}$$

As with point sources, a further simplification for ground level receptors, for which z = 0, yields:

$$C = \frac{2Q}{\sqrt{2\pi}\,\sigma_z u}\left\{\exp\left[-\frac{h^2}{2\sigma_z^2}\right]\right\}$$

Where both source and receptor are at ground level, h = z = 0 making the exponential terms unity. Hence

$$C = \frac{2Q}{\sqrt{2\pi}\,\sigma_z u}$$

The absence of the horizontal dispersion parameter, σ_y, is explained by the assumption that lateral dispersion in one segment of the line is compensated by dispersion in the opposite direction from adjacent segments. Additional equations for line sources are given elsewhere.* Clearly, when modelling without the aid of a computer, it is important to be able to identify which equation to use for a particular application. Exercise 5 in Chapter 6 demonstrates the important difference between elevated and ground level sources in terms of their impact at ground level, and requires the use of the appropriate modelling equations. However, before applying any of the equations to practical problems, the terms that go into these equations must be examined in more detail.

* Middleton, D R, 1997, *Manual on Modelling: A Guide for Local Authorities*, Met O (APR) Turbulence and Diffusion Note No 241.

The Gaussian Dispersion Coefficients, σ_y and σ_z

Critical input parameters to the Gaussian model are these standard deviations (σ) along the axes. Also known as dispersion coefficients, these values express the spread of concentrations across the plume dimensions. Normally, the values of σ_y and σ_z are determined most simply from knowledge of the atmospheric stability. For a given stability class, the diffusion coefficient can be stated as a function of the downwind distance in metres, using the methods described later in this section. It should be noted that these coefficients represent an atmosphere without an inversion barrier. The question is how do we identify the atmospheric stability? The answer to that was given in Chapter 3, which also described the influence of urban development on atmospheric stability, since this has an influence on the dispersion coefficients. As the numerical values of the diffusion (or spreading) coefficients are critical to the application of the Gaussian forms and since there is no universal method of achieving them, their evaluation has been discussed at length in the literature.

The values may be obtained by either theoretical or empirical approaches. The scheme proposed by Pasquill in 1961 and described in Chapter 3 involves subjective estimates of the structure of turbulence in the atmospheric boundary layer. Pasquill obtained the vertical and the lateral spreading from routine meteorological observations consisting of wind speed, insolation and cloudiness. He presented this information as functions of the six atmospheric stability classes identified in Table 3.7. Gifford modified the scheme somewhat using families of curves of σ_y and σ_z respectively. These are called the Pasquill–Gifford curves. They are based on wind speeds > 2 ms^{-1} with a sampling time of three minutes, and so may be conservative. Knowing the atmospheric stability, one of the simplest ways to determine the appropriate value of the dispersion coefficient at a given distance is by interpolation from the relevant graph (Figures 5.4 and 5.5).

These graphs were derived from the equations in Table 5.1. Values for σ_x are not included since it is assumed that $\sigma_x = \sigma_y$. Values for σ_x become important in considering releases from emergency releases and unplanned events, and such situations will be considered in more detail in Chapter 8.

Table 5.1 Equations and Data for Pasquill–Gifford Dispersion Equations Applied to Continuous Plumes

Pasquill Stability class	σ_y (m)	Downwind distance x (m)	σ_z (m)
A	$\sigma_y = 0.493x^{0.88}$	100–300	$\sigma_z = 0.087x^{1.10}$
		300–3000	$\log_{10}\sigma_z = -1.67 + 0.902\log_{10}x + 0.181(\log_{10}x)^2$
B	$\sigma_y = 0.337x^{0.88}$	100–500	$\sigma_z = 0.135x^{0.95}$
		500–20 000	$\log_{10}\sigma_z = -1.25 + 1.09\log_{10}x + 0.0018(\log_{10}x)^2$
C	$\sigma_y = 0.195x^{0.90}$	100–100 000	$\sigma_z = 0.112x^{0.91}$
D	$\sigma_y = 0.128x^{0.90}$	100–500	$\sigma_z = 0.093x^{0.85}$
		500–100 000	$\log_{10}\sigma_z = -1.22 + 1.08\log_{10}x + 0.061 (\log_{10}x)^2$
E	$\sigma_y = 0.091x^{0.91}$	100–500	$\sigma_z = 0.082x^{0.82}$
		500–100 000	$\log_{10}\sigma_z = -1.19 + 1.04\log_{10}x + 0.070(\log_{10}x)^2$
F	$\sigma_y = 0.067x^{0.90}$	100–500	$\sigma_z = 0.057x^{0.80}$
		500–100 000	$\log_{10}\sigma_z = -1.91 + 1.37\log_{10}x + 0.119(\log_{10}x)^2$

Figure 5.4 Dispersion Coefficients Along the y-Axis (σ_y)

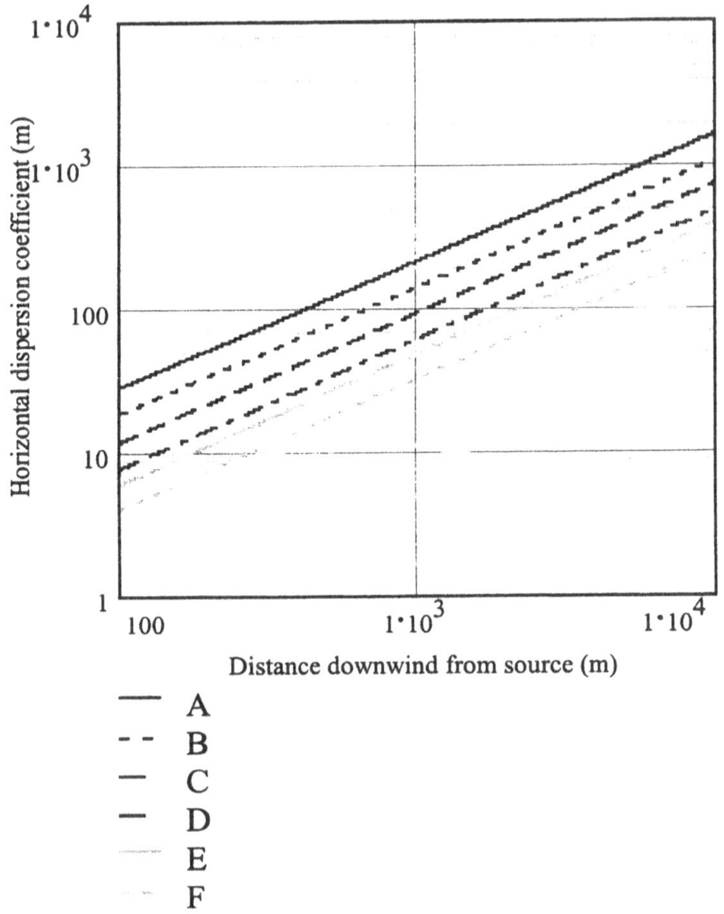

Figure 5.5 Dispersion Coefficients Along the z–Axis (σ_z)

(a) distances to 500 m

(b) distances beyond 500 m

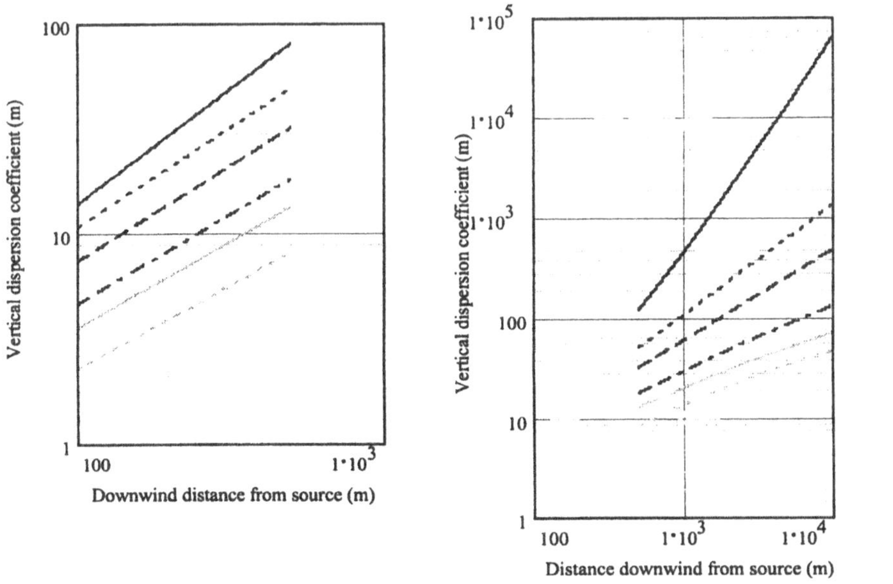

Table 5.2 Data Applied to Puff Releases

Stability	**x = 100 m**		**x = 4000 m**	
	σ_y (m)	σ_z (m)	σ_y (m)	σ_z (m)
very stable	1.3	0.75	35	7
neutral	4	3.8	120	50
unstable	10	15	300	220

Alternative equations have been derived empirically for a given situation in an attempt to match predicted and measured concentrations. It will be recalled from Chapter 3 that the urban effect can shift the stability category. An approach proposed by Briggs takes account of differences in urban and open-country conditions. Briggs proposed a series of interpolation formulae for the σ curves. They apply to plumes from elevated stack sources, so that they reflect diffusion data for a higher source at greater downwind distances. The equations for these curves are summarized in Table 5.3. While the Briggs functions are assumed to be valid for downwind distances (x) less than 100 m, concentrations closer than 100 m from a source may be suspect.

Table 5.3 Formulae Recommended by Briggs for σ_y and σ_z (100m < x <10000m)

	σ_y	σ_z
Open-country Conditions		
A	$0.22x(1+0.0001x)^{-0.5}$	$0.20x$
B	$0.16x(1+0.0001x)^{-0.5}$	$0.12x$
C	$0.11x(1+0.0001x)^{-0.5}$	$0.08x(1+0.0002x)^{-0.5}$
D	$0.08x(1+0.0001x)^{-0.5}$	$0.06x(1+0.0015x)^{-0.5}$
E	$0.06x(1+0.0001x)^{-0.5}$	$0.03x(1+0.0003\,x)^{-1}$
F	$0\ 04x(1+0.0001x)^{-0.5}$	$0.016x(1+0.0003x)^{-1}$
Urban Conditions		
A–B	$0.32x(1+0.0004x)^{-0.5}$	$0.24x(1+0.001x)^{-0.5}$
C	$0.22x(1+0.0004x)^{-0.5}$	$0.20x$
D	$0.16x(1+0.0004x)^{-0.5}$	$0.14x(1+0.0003x)^{-0.5}$
E-F	$0.11x(1+0.0004x)^{-0.5}$	$0.08x(1+0.00015x)^{-0.5}$

It is possible for the stability class estimate inferred from the observed dispersion to differ by two or more classes from the estimate based on meteorology. The observed horizontal and vertical dispersion classes can differ from one another by two or more classes. On the other hand, it has been observed that the Pasquill–Gifford dispersion curves with the stability category determined from wind direction fluctuation performed well in urban studies, whereas the Briggs dispersion curves resulted in considerable under calculation, especially at night.

Irwin extensively reviewed sigma schemes and noted that the applicability of a particular scheme is dependent on source height, surrounding roughness and stability. Many studies estimate diffusion based on the simpler Pasquill stability classes because they are easier to use and tend to give satisfactory results. More importantly, the instruments, time and expertise required to gain higher quality turbulence observations cannot always be justified. Nevertheless, it must always be borne in mind that the sigma curves based on the Pasquill classes are experimental observations and only strictly applicable to the area of their derivation. In other words, they are not appropriate in instances such as complex terrain and for distances greater than 10 km, or for effective release heights of above 100 m. In such cases either direct turbulence measurements or theoretical extrapolations are necessary. Although such experiments highlight the difficulty of specifying the turbulent mixing within the atmosphere, the analysis of dispersion experiments is also dependent on the assumptions of conservation of tracer mass (that is no loss of the chemical tracer through deposition, chemical reaction, etc) and an assumed vertical profile. It was observed in the Prairie Grass experiment that the deposition and vegetative uptake of sulphur dioxide, led to a 20–25%

reduction in the ground level concentration at 200 m from the source. Hence, failure to account for this pollutant loss would imply larger values of σ than in reality.

The method of Pasquill is followed with frequent warnings to limit it to specific conditions and similar caveats. It will be recalled that Pasquill defined categories of atmospheric stability in terms of readily measured parameters, such as surface wind, cloudiness (at night), or solar radiation (daytime). The spreading coefficients are defined for a given sampling interval and the lower few hundred metres of the atmosphere. Many practical problems relate to the estimation of plume spread coefficients. On clear, calm nights the stable air allows the slow meandering of plumes leading to cross-wind spread greater than expected. Equally, there may be problems when cool air drains into valleys. To overcome such problems it is possible to use a bivane to record the horizontal and vertical wind directions and thence their standard deviations. Equations are available to derive dispersion coefficients from such measurements. Ultrasonic anemometers are increasingly used nowadays for measuring winds. Modern models may use wind measurements as input data directly.

The selection of either rural or urban dispersion coefficients in a specific application may follow a land use classification procedure or a population based procedure to determine whether the character of an area is primarily urban or rural. Conventional US land use procedures are as follows:

- Classify the land use within the total area, A_o, circumscribed by a 3 km radius circle about the source using the meteorological land use typing scheme. If land use types I1, I2, C1, R2, and R3 account for 50% or more of A_o, use urban dispersion coefficients; otherwise, use appropriate rural dispersion coefficients.
- Alternatively, using the population density procedure, compute the average population density, p per square km with A_o as defined above. If p is greater than 750 people km^{-2}, use urban dispersion coefficients; otherwise use appropriate rural dispersion coefficients.

Of the two methods, the land use procedure is considered more definitive and has been applied more widely. A more detailed description of the procedure is given in Appendix 2. Population density should be used with caution and should not be applied to highly industrialized areas. This is because the population density may be low and so while a rural classification would be indicated, the area is sufficiently built-up to satisfy the urban land use criteria. In this case, the classification should already be 'urban' and urban dispersion parameters should be used. Inevitably there will be problems, such as when a factory is located on the edge of town. It may be rural on one side and urban on the other. Sea breeze effects also present special problems and the latest generation of models can deal with these.

The Effective Stack Height, H

It is common for releases into the atmosphere from stacks (or chimneys) to have a temperature above that of their surroundings. In addition, releases usually have an initial vertical momentum because of a fan within the system. Gases leaving the stack rise when they are either of a lower density than the surrounding air (buoyancy rise) or are ejected at a velocity high enough to give the exit gases upward kinetic energy (momentum rise). Buoyancy rise is sometimes called thermal rise because the most common cause of lower density is higher temperature. The thermal buoyancy and the vertical momentum combine to result in the plume seeming to result from a source at a greater effective height than the physical height of the stack. The effective height (H) is therefore the sum of the physical stack height (h) and the plume rise (ΔH) as suggested in Figure 5.6. Exceptions are for emissions of gases that have higher density than the surrounding air, or when stack downwash occurs.

The effective stack height is the height at which the plume becomes level, and this may not occur until some considerable distance from the stack. Indeed, it may be difficult to determine when the plume becomes level, as Figure 5.6 suggests.

Figure 5.6 The Effective Height is Influenced by Plume Rise

An important input parameter for a dispersion model is this effective stack height. It is the height at which the plume essentially becomes level, and it will rarely be the same as the physical height of the stack. In fact, plume rise can increase the effective stack height by a factor of twice to ten times the physical stack height and so is an important factor in determining ground level concentrations. Plume rise may be due to gases leaving the top of the stack being less dense than the surrounding air (buoyancy rise) or because they are ejected at a high enough velocity to give momentum rise. Many empirical equations have been derived for estimating plume rise. Early studies based on regression equations produced empirical equations that are of limited applicability. Since the plume rise from stacks generally occurs over some distance downwind these equations should not be applied within the first few hundred metres of the stack. In an assessment of potential impact of a gas turbine power station, several of these equations were compared.* One point that was clear from this study is the variety of equations, which are not directly comparable in output, with predicted plume rise ranging from 72 m to 330 m. It was also evident that for a stack height of 55 m, the effective height of the plume is double this within one stack distance from the source. Furthermore, most of these semi-empirical formulae follow a general form:

$$\Delta h = KQ_h^{a}\, x^{b}\, u^{c}$$

where:

K = a constant depending on variables that influence h — Q_h = the heat emission rate

x = the downwind distance from the stack — u = the wind speed

a, b, c = powers derived from theory or experiment

Several studies have suggested values of $\frac{1}{3}$, $\frac{2}{3}$ and –1 respectively.

The choice of the most appropriate plume rise equation is particularly important in calculations for regulatory purposes.

Briggs made the fundamental break from the traditional approach and of the variety of equations available, his are widely used. In the Briggs Model, the starting point is to get a measure of the buoyancy of an emission by defining the buoyancy flux parameter (F). If the effluent has the same molecular mass and specific heat as air, this value is defined as:

$$F = g\nu r^2 \left(\frac{T_s - T}{T_s}\right)$$

where:

g = acceleration due to gravity (9.8 m s^{-2}) — ν = efflux velocity from the stack (m s^{-1})

r = inside radius of stack (m) — T_s = stack gas exit temperature (K)

T = ambient temperature (K)

The units of F are $m^4\ s^{-3}$. This expression gives a reasonable approximation for most applications.

* HMIP, *An Assessment of the Effects of Industrial Releases of Nitrogen Oxides in the East Thames Corridor*, HMIP Environment series, HMSO, London, 1993

Working equations often distinguish between stability conditions. A starting point is to determine the distance (x_f m) to final plume rise. Under neutral or unstable conditions in the atmosphere (Pasquill stability categories A to D), values of x_f and then plume rise (Δh) are derived as follows:

If $F \geq 55\ m^4\ s^{-3}$ $\qquad x_f = 119F^{0.4}$ $\qquad \Delta h = 38.71\left(\frac{F^{0.6}}{u_s}\right)$

If $F < 55\ m^4\ s^{-3}$ $\qquad x_f = 49F^{5/8}$ $\qquad \Delta h = 21.425\left(\frac{F^{0.75}}{u_s}\right)$

The value u_s is the wind speed adjusted to the release height.

For stable conditions (Pasquill stability categories E and F), it is necessary to calculate the stability parameter (s). The value of s is given by:

$$s = \frac{g}{T}\left(\frac{dT}{dz} + \Gamma\right)$$

where g is the acceleration due to gravity and Γ is the adiabatic lapse rate, 0.01 K m^{-1}. Here, s is the square of what is known as the Brunt–Vaisala frequency, which related to the oscillating motion that may be set up when buoyancy forces operate to restore the displacement forces on a parcel of air. It is significant where vertical oscillations of the air may bring pollutants to the ground.

The derivative $\frac{dT}{dz}$ represents the actual rate of change of temperature with altitude. Where this is not available, a typical value of the term within parentheses for stability E is 0.02 K m^{-1}, while for stability F a value of 0.035 K m^{-1} may be used. Then the plume rise is determined as follows.

When there is wind: $\qquad x_f = 2.0715\left(\frac{u_s}{\sqrt{s}}\right)$ $\qquad \Delta h = 2.6\left(\frac{F}{su_s}\right)$

For calm conditions: $\qquad \Delta h = 4\left(\frac{F^{0.25}}{s^{0.375}}\right)$

Under stable conditions, the lower value of these two estimates is taken as the plume rise.

If the stack temperature is below or only slightly above ambient temperature, then the plume rise due to momentum dominates. For unstable and neutral conditions:

$$\Delta h = 6r\frac{\upsilon}{u_s}$$

Briggs suggested that this equation was most applicable when (ν/u_s) exceeds 4.

For low buoyancy plumes in stable conditions, the plume rise is estimated from:

$$\Delta h = 1.5\left(\frac{\upsilon r^2 T_a}{T_s u_s \sqrt{s}}\right)^{1/3}$$

This value should be calculated and compared with that from the preceding equation, and the lower value of the two used in modelling.

An example of steps taken to address the known deficiencies in existing methods for modelling plume rise and building-induced downwash is research sponsored by the Electric Power Research Institute (EPRI) and a consortium of electric utilities in the US. The research integrated field observations and laboratory fluid modelling simulations with numerical and analytical modelling to develop the Plume Rise Model Enhancements-PRIME.

The field study was conducted in 1994 at Jersey Central Power & Light's Sayreville Power Plant in New Jersey, US. The company monitored unit operations and emission characteristics on a continuous basis as plant load was varied over a substantial range. Meteorological parameters were also monitored using a limited number of tethered balloon soundings to supplement ground-based instruments.

Plume concentrations were mapped three-dimensionally by means of mobile aerosol backscatter lidar that measured the concentration of fine smoke particles within the plume. Concentrations in the building cavity were measured by a series of monitors for nitrogen oxides. Supplementing the field measurements were several sets of wind tunnel studies.

The physical processes included in PRIME include the perturbation to wind flow streamlines caused by the presence of an obstructing structure. A plume's downwash trajectory is simulated using an empirical relationship that varies with the shape of the building and the wind angle, coupled with a numerical plume rise formulation. This approach allows PRIME to determine the change in plume centre line location with downwind distance. PRIME is incorporated in the latest generation of dispersion models including ISC.

Model Calculation

To illustrate the use of an appropriate modelling equation for calculating the impact of a release, consider a stack with an effective height of 100 m emitting sulphur dioxide at a rate of 8 g s^{-1}. It is required to predict the ground level concentration 3 km downwind and directly under the plume. In the UK, stability D predominates and from Figures 5.4 and 5.5, the dispersion coefficients are found to have values $\sigma_y = 180$ m and $\sigma_z = 67$ m. The wind speed is 5 m s^{-1}. Substituting in the equation:

$$C = \frac{8}{\pi \times 5 \times 180 \times 67} \exp\left(-\frac{100^2}{2 \times 67^2}\right) = 4 \times 10^{-5} \text{ g m}^{-3}$$

Such calculations can be repeated at different distances from the source, but it becomes tedious to do this by hand. However, a computer makes this very easy to do and can produce a graph showing the variations in ground level concentration with distance, as Figure 5.7 illustrates.

Figure 5. 7 Variation of Ground Level Concentration with Distance from an Elevated Source of Pollution

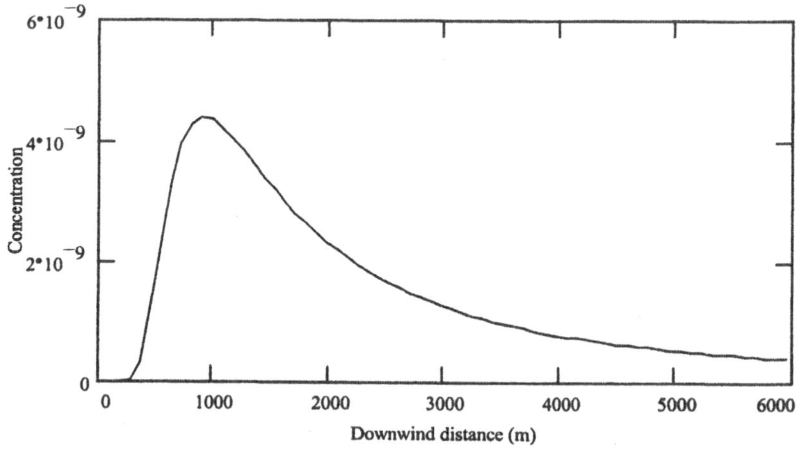

These repetitive calculations may be taken further to test the sensitivity of the impact at ground level to changes in atmospheric conditions such as different stabilities and at other locations.

The so-called R91 model produced by a working group of the National Radiological Protection Board in the UK was based on the Gaussian plume concept, and for many years was the closest approach to UK standard practice. This model uses many of the fundamental principles and equations introduced previously. The use of the equations and nomograms developed from them are the simplest approach to dispersion modelling and should be the first approach in any application. Many eulerian dispersion models based on the Gaussian plume concept have been, and continue to be developed, and those from the US EPA have assumed a certain 'regulatory' status. These models are for use on a computer and are available for purchase through various intermediaries, or can be downloaded from the Technology Transfer Network via the Internet. Such models vary in complexity from simple models for screening through to more advanced models.

Typically a tiered approach can be used to refine the estimate of impact of a stationary source releasing a contaminant to atmosphere. This purpose of this is to minimize the time and costs of producing information. The stages are as follows:

- Start with a simple model such as R91. If this shows that concentrations are only a small fraction of a relevant limit concentration, then the project may be regarded as acceptable without further modelling. The fraction of the standard should be based on the accuracy of the concentration estimates together with a safety factor.
- If the simple model suggests an unacceptable level, then the next level of model is appropriate. This could be a computer model such as SCREEN or ISC described later.

Evidence of a problem with intermediate modelling suggests application of an advanced model such as AERMOD or ADMS. Such models consider the atmospheric boundary layer, not in terms of a single Pasquill category, but take account of variations through the boundary layer depth. In addition, ADMS models dispersion under convective meteorological conditions using a skewed Gaussian concentrations distribution, shown by validation studies to be a better representation than a Gaussian expression. More detail of these approaches will be developed in the following chapters.

To summarize the major differences between each of the three modelling stages, Table 5.4 briefly lists the input requirements, output parameters, and assumptions associated with each stage. Risk assessments such as cancer unit risk estimates, chronic non-cancer concentration thresholds, and acute concentration thresholds may be used to convert concentration predictions into cancer risks, chronic non-cancer risks, and acute non-cancer risks, respectively.

Table 5.4 Stages for Modelling

Modelling stage	Input requirements	Output parameters	Major assumptions
Stage 1	Emission rate, stack height, meteorological data.	Maximum off site concentrations, comparison against air quality standards, worst-case cancer risk or worst-case non-cancer risk (short and long term).	Worst-case meteorology, worst-case downwash, worst-case stack parameters, short term releases occur simultaneously, maximum impacts co-located, cancer risks additive, non-cancer risks additive.
Stage 2	Emission rate, stack height, efflux velocity, stack temperature, stack diameter, rural/urban site classification, building dimensions for downwash calculation.	Maximum offsite concentrations, comparison against air quality standards, worst-case cancer risk and/or worst-case non-cancer risk (short and long term).	Worst-case meteorology, short term releases occur simultaneously, maximum impacts co-located, cancer risks additive, non-cancer risks additive.
Stage 3	Emission rate, stack height, release point locations, efflux velocity, stack temperature, stack diameter, rural/urban site classification, terrain, local meteorological data, receptor locations for concentration predictions, frequency and duration of short term (intermittent) releases.	Concentrations at each receptor point, comparison against air quality standards, long term cancer risk estimates, chronic non-cancer risks at each receptor point.	Cancer risks additive, non-cancer risks additive.

A tiered approach is suggested in the guidance for local authorities in producing an air quality action plan.* This suggests that it will probably not be practicable for local authorities to use complex modelling software to assess a wide range of options to reduce air pollution and, therefore, a relatively simple screening approach could be used to assemble option packages.

Conclusion

The most commonly used models are probably Gaussian plume dispersion models that predict ground level concentrations of conservative pollutants within a few kilometres of the source. Starting from a three-dimensional axis system of downwind, crosswind, and vertical coordinates with the origin at the ground, these models make several assumptions. First, that concentrations from a continuously emitting plume are proportional to the emission rate. Second, that these concentrations are diluted by the wind at the point of emission at a rate inversely proportional to the wind speed; and that the time-averaged pollutant concentrations crosswind and vertically near the source are well described by Gaussian or normal (bell-shaped) distributions. The standard deviations of plume concentration in these two directions are empirically related to the levels of turbulence in the atmosphere and increase with distance from the source. In their simplest form, Gaussian models assume that the pollutant does not undergo chemical reactions or other removal processes in travelling away from the source. It is also assumed that pollutant material reaching the ground or the top of the mixing height as the plume grows is eddy-reflected back inward the plume centre line.

The unresolvable component of wind speed includes turbulent atmospheric diffusion eddies that have an effect orders of magnitude greater than molecular diffusion. Appreciating that uncertainties in such components cannot be eliminated is a key factor in understanding the fundamental difference between modelled and measured air quality data. Turbulent eddies cause inevitable and unresolvable variations in wind speed at a given point and time. These variations in wind speed leads to an inevitable uncertainty in concentration predicted at a given point and time. Various expressions for the dispersion coefficients have been derived, but differences lead to more uncertainties. Measured air quality, however, is the actual concentration subject only to the errors of the measurement technique. Hence, there is a conceptual difference between measured air quality and modelled data because of these intrinsic uncertainties. All of this assumes that the information on emissions etc put into the model is good. To quote a well-known phrase 'rubbish in, rubbish out'. The overall conclusion is that modelled data must never be regarded as absolute, nor must they be regarded as very accurate, despite the huge computing power and large number of significant figures in the results generated.

Despite their limitations, Gaussian models have enjoyed a wide degree of popularity mainly due to their simplicity and the fact that they give reasonable results. Gaussian plume dispersion models can also be formulated to suit line sources (for example roads including networks of roads); area sources (for example towns); mobile sources (for example a ship); complex sources (for example airports) as well as conventional point sources (for example single stacks). Special dispersion models are also available for the behaviour of dense gases, usually applied to accidental release (for example chlorine). While predictions may not be accurate, modelling is a useful tool if these cautionary notes are borne in mind. The latest generation of models such as ADMS and AERMOD use a more advanced understanding of the atmosphere to improve predictions.

* *Air Quality Action Plans: Interim Guidance for Local Authorities*, National Society for Clean Air and Environmental Protection, Brighton.

Exercises

1. A manufacturing organization located on the coast near a proposed hospital site releases combustion products from its power plant into the atmosphere through three stacks as follows.
   Stack 1 0.5 tonnes per day released at 74.5 m.
   Stack 2 0.75 tonnes per day released at 100 m.
   Stack 3 0.25 tonnes per day released at 102 m.
   The locations of the sources are indicated in Figure 5.8.

   Figure 5.8 Scenario for Question 1

   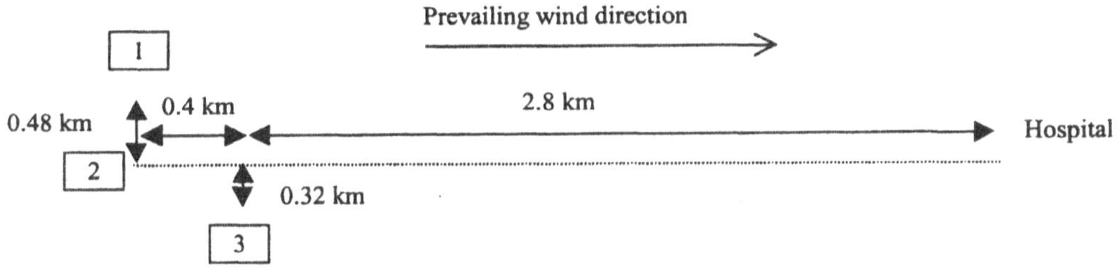

   Under the prevailing wind conditions, the factory is generally upwind of the proposed hospital, which will be a single storey development. Sulphur dioxide is a respiratory irritant and causes tightening of the airways when inhaled at high concentrations. Those suffering from asthma are significantly more sensitive to sulphur dioxide than other people. There is little evidence to suggest that those suffering from asthma would be significantly affected by exposure to concentrations of sulphur dioxide of less than 200 ppb. The WHO has suggested that exposure to 400 ppb sulphur dioxide may lead to significant narrowing of the airways in those suffering from asthma. For most people the effects expected would not be large, although some individuals may be clinically affected. Exposure to such concentrations may add to the effects of exposure to other pollutants and allergens. Consequently, asthmatics may need to increase their medication since the effects can be reversed by the use of inhalers. As concentrations rise above 400 ppb then more asthmatic individuals may experience adverse effects. Determine whether the stack emissions will be of concern in terms of air quality at the proposed hospital site under a worst case scenario.
2. Estimate the plume rise for an emission from a 30 m stack of diameter 1 m on a warm sunny day when the ambient temperature is 20°C and the pressure 1013 mb, with a wind speed at the top of the stack being 3 m s^{-1}. The efflux velocity of the emission is 20 m s^{-1} and its temperature is 400 K.
3. An electricity generating station has two coal-fired boilers each rated at 31.5 MW_{th} and venting through a 77 m stack. The exit velocity of the gases is 8.8 ms^{-1} at a temperature of 140°C. Estimate the ultimate effective height of the stack under stable atmospheric conditions where the atmospheric temperature increases at a rate of 2°C km^{-1}. Assume a stack diameter of 4 m, an ambient temperature of 20°C and winds at the stack height are 5 m s^{-1}.

CHAPTER 6
SIMPLE MODELLING FOR POINT AND MOBILE SOURCES

The aim of this chapter is to introduce simple techniques for air quality modelling. Perhaps the simplest approach is to calculate concentrations at specific points using an appropriate equation from Chapter 5. This may be done by hand or using a calculator or computer. There are also workbook approaches to modelling.

Screening

A first stage screening technique can be carried out using direct calculation, workbook or graphical procedures in order to identify the likely significance of sources, and, perhaps, to eliminate them from further consideration. Workbook procedures can also deal with elevated point sources with plume heights between 10 m and 300 m, and can yield information on concentrations averaged over 1 hr to 24 hrs. The techniques assume that there is no aerodynamic downwash. There are also simple computer programmes for first stage screening based on the Gaussian models. All three approaches will be outlined in this chapter.

The second stage for assessing air quality impacts should be used in those cases where a simple screening (first phase) procedure indicates that an air quality problem may occur. Such procedures will be introduced in Chapter 7.

Direct calculation

Direct calculation from the basic Gaussian plume equations can be used as perhaps the simplest approach to modelling and one example was given in Chapter 5. The use of mathematical software such as Mathcad® can simplify this if several calculations have to be used (Figure 6.1).

Figure 6.1 Mathcad® Programmed for Dispersion Calculations

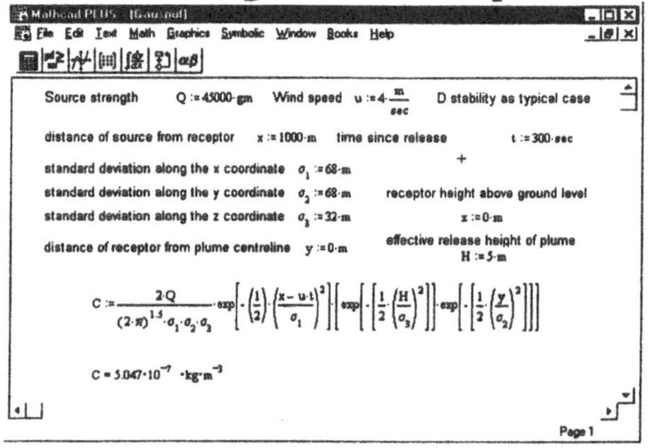

The graphs for determining values of the dispersion coefficients (Figures 5.4 and 5.5) in Chapter 5 were generated in Mathcad®. Similarly, plotting curves of the type in Figures 5.7 illustrates the practicality of this approach if a microcomputer is used for the repetitive calculations. It is also possible to programme the equations into a spreadsheet such as Excel. Workbook methods such as the Design Manual for Roads and Bridges are available in this form.* A toolkit that is available on the Internet is represented in Figure 6.2.

Figure 6.2 Dispersion Modelling on the Internet (www.industrialhygiene.com/)

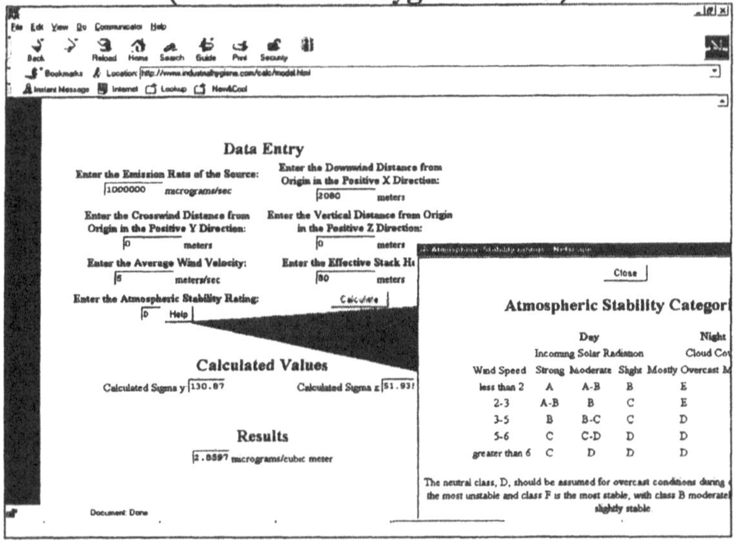

Where is the Maximal Ground Level Concentration?

One outcome of common concern is the location of the highest concentration at ground level. This may be needed for assessing environmental effects or for identifying where to place monitoring equipment. Traditional wisdom as related in Chapter 2 suggests that the maximum annual mean concentration occurs about 15 stack heights from the source. However, a simple graphical approach is also available for enabling such calculations to be done by hand.† A compilation of the data for the various conditions is presented in Figures 6.3(a) and 6.3(b).

The series of lines relate effective stack height, atmospheric stability and maximal concentration. To avoid repetitious computations to calculate ground level concentrations, values of the product of relative ground level concentration and wind speed have been calculated at various downwind distances for various effective emission heights and stability categories. This product is conventionally denoted as $\chi u/Q$ where χ represents the ground level concentration, Q represents the source emission rate and u is the wind speed. The actual ground level concentration may be determined by multiplying the value of $\chi u/Q$ by Q/u.

Knowing the effective emission height and atmospheric stability, the position of the maximum ground level concentration and the value of $\chi u/Q$ can be read from the appropriate Figure 6.3. Multiplying $\chi u/Q$ by Q/u gives the value for the maximum ground level concentration. This approach forms part of a simple screening procedure that is the 'first phase' recommended when assessing the air quality impact of a new point source under US legislation. The purpose of this screening procedure is to eliminate from further analysis those sources that clearly will not cause or contribute to ambient concentrations that exceed short term air quality standards. The scope of the procedure is confined to elevated point sources with plume heights of 10 m to 300 m, and concentration averaging times of 1 hr to annual. When applying the screening procedure to elevated point sources, no aerodynamic downwash of the effluent plume by nearby buildings should occur, nor should the plume impact on elevated terrain. If the potential for building downwash exists, then

* http://www.stanger.co.uk/airqual/modelhlp/dmrb/dmrb.asp

† Turner D B, *Workbook of Atmospheric Dispersion Estimates*, First Edition, US Public Health Service Publication No. 26, 1970, Second Edition 1994, Lewis.

a simple computer model such as SCREEN should be used in the first instance to estimate air quality impact as described later in the chapter.

Figure 6.3(a) Downwind Distance to Maximum Concentration and Maximum Xu/Q as a Function of Stability Class for Rural Terrain

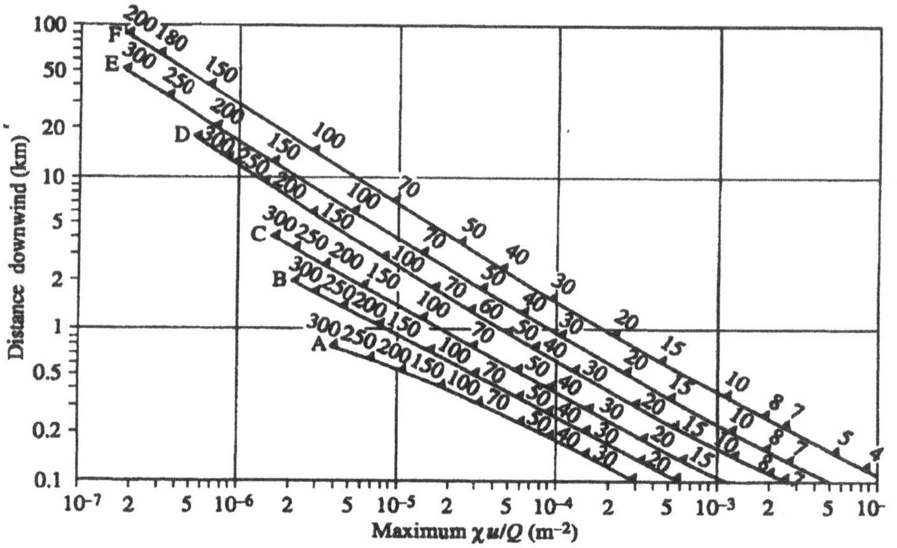

Note: Plume heights (m) are shown on the lines.

Figure 6.3(b) Downwind Distance to Maximum Concentration and Maximum Xu/Q as a Function of Stability Class for Urban Terrain

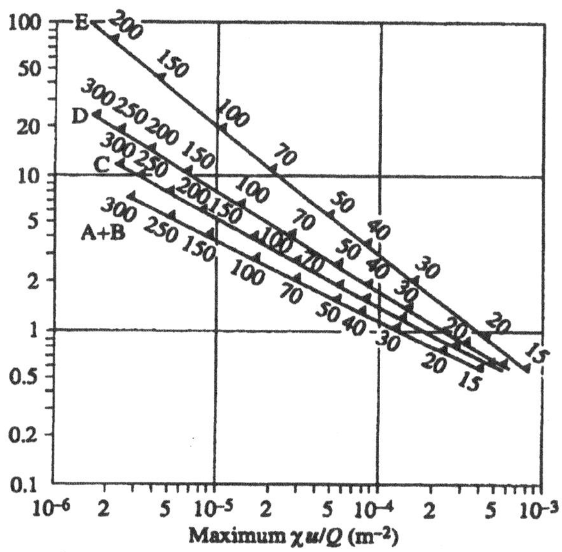

Note: Plume heights (m) are shown on the lines.

The effects of elevated terrain below stack height should also be accounted for by reducing the computed plume heights by the maximum terrain height above stack base. The screening procedure uses the Gaussian dispersion equation to estimate the maximum 1 hr ground level concentration for the source in question. To obtain concentrations for other averaging times up to annual, it is necessary to multiply the 1 h value by an appropriate factor (Chapter 3). Background concentrations must also be considered to obtain a total concentration estimate, which is used to determine if further analysis of the impact is warranted. A simple approach involves determining the effective release height and then using Figure 6.3 to estimate a value of $\chi u/Q$. Divide the $\chi u/Q$ value by the wind speed to determine the corresponding χ/Q value

$$\frac{\chi}{Q} = \frac{\chi u/Q}{u}$$

Then multiply the maximum χ/Q value by the emission rate Q (g s^{-1}). Incorporate a factor of two margin of safety, to obtain the maximum 1 hr ground level concentration $\chi 1$ (g m^{-3}) due to emissions from the stack in question:

$$\chi_1 = 2Q\left[\frac{\chi}{Q}\right]$$

The margin of safety is incorporated in the screening procedure to account for the potential inaccuracy of concentration estimates obtained through calculations of this type. To obtain a concentration estimate (χ_p) for an averaging time greater than 1 hr, multiply the 1 hr value by an appropriate factor, r:

$$\chi_p = r\,\chi_1$$

Next, contributions from other sources to the background concentration (χ_B) should be taken into account, giving the final estimate χ_{max} (g m^{-3}):

$$\chi_{max} = \chi_p + \chi_B$$

Based on the estimate of χ_{max} and local circumstances such as impaction on elevated terrain, it may be determined whether further analysis of the source is justified. If the estimated concentration is below a level of concern (for example an air quality standard), the source can safely be assumed to pose no threat to that air quality level, and no further analysis is necessary.

This procedure is relatively simple, and may be applied without the need for computing resources. However, care should be taken under certain weather conditions. For example, recall from Figures 3.7(b) and 3.8 that under very unstable conditions a looping plume may reach the ground level relatively close to the source, and this may lead to high concentrations.

Another example of a phased approach starting with a simple calculation procedure has been devised for sources of toxic substances.* Since many releases of hazardous air pollutants are intermittent (for example batch process emissions), techniques have to address intermittent as well as continuous sources. It is important to understand the different treatment of emission rates for both types of sources when carrying out either the analysis of a long term impact or a short term impact. In long term impact analysis, the emission rate used for modelling is based on the amount of pollutant emitted over a 1 yr period, regardless of whether the emission process is continuous or intermittent. In addition, to assess the worst-case impact of a source or group of sources, long term emission rates used in model simulations should reflect the emission rates for a plant or process that is operating at full design capacity. In short term impact analysis, the emission rate used for modelling is based on the maximum amount of pollutant emitted over a 1 hr period, during which time the release occurs. Long term modelling estimates pollutant concentrations as annual averages. Usually the assessment is concerned with impact outside the boundary of an industrial site. The reference cited should be consulted for details of the methodology.

Computer Screening Models

The preliminary assessment of an air pollution problem can be accomplished by the use of screening models that are designed to determine quickly and easily the impacts from a single source. Screening is not appropriate if it is apparent that several sources contribute to ambient concentrations. An example of a screening model is SCREEN.† The following information box shows a typical SCREEN run.

---

* EPA-450/4-92-001 *A tiered modelling approach for assessing the risks due to sources of hazardous air pollutants.*

† EPA 450/4-9-006, *Modelling Guideline* and EPA 454/R-92-019 *Screening Procedures for Stationary Sources.*

```
25/08/01                                                                    16:03:28
 ***  SCREEN MODEL RUN  ***
 *** VERSION DATED 92245 ***
SIMPLE TERRAIN INPUTS:
   SOURCE TYPE                =        POINT
   EMISSION RATE (G/S)        =          .420000
   STACK HEIGHT (M)           =         40.0000
   STK INSIDE DIAM (M)        =           .5000
   STK EXIT VELOCITY (M/S)    =          5.6000
   STK GAS EXIT TEMP (K)      =        303.0000
   AMBIENT AIR TEMP (K)       =        293.0000
   RECEPTOR HEIGHT (M)        =           .0000
   URBAN/RURAL OPTION         =        URBAN
   BUILDING HEIGHT (M)        =          4.0000
   MIN HORIZ BLDG DIM (M)     =          5.0000
   MAX HORIZ BLDG DIM (M)     =         10.0000
BUOY. FLUX =  .113 M**4/S**3;  MOM. FLUX =  1.895 M**4/S**2.
 *** STABILITY CLASS  4 ONLY ***
 ***********************************
 *** SCREEN AUTOMATED DISTANCES ***
 ***********************************
 *** TERRAIN HEIGHT OF  0. M ABOVE STACK BASE USED FOR FOLLOWING DISTANCES ***
DIST   CONC              U10M   USTK   MIX HT   PLUME    SIGMA    SIGMA
 M)  (UG/M**3)   STAB   (M/S)  (M/S)    (M)     HT (M)   Y (M)    Z (M)   DWASH
------ ----------  ----  -----  -----   ------   -------  -------  ------  -----
50     .1719E-04    4     8.0    11.3   2560.0    39.74    7.92     6.95    NO
100  1.867          4     1.5     2.1    480.0    43.96   15.73    13.84    NO
200  27.16          4     1.0     1.4    320.0    45.94   30.84    27.25    NO
 MAXIMUM 1-HR CONCENTRATION AT OR BEYOND    50. M:
241. 29.17          4     1.0     1.4    320.0    45.94   37.01    32.76    NO
  DWASH=       MEANS NO CALC MADE (CONC = 0.0)
  DWASH=NO     MEANS NO BUILDING DOWNWASH USED
  DWASH=HS     MEANS HUBER-SNYDER DOWNWASH USED
  DWASH=SS     MEANS SCHULMAN-SCIRE DOWNWASH USED
  DWASH=NA     MEANS DOWNWASH NOT APPLICABLE, X<3*LB
 *** CAVITY CALCULATION - 1 ***       *** CAVITY CALCULATION - 2 ***
  CONC (UG/M**3)        =   .0000       CONC (UG/M**3)          =      .0000
  CRIT WS @10M (M/S) =  99.99           CRIT WS @10M (M/S)      =   99.99
  CRIT WS @ HS (M/S)  =  99.99          CRIT WS @ HS (M/S)      =   99.99
  DILUTION WS (M/S)   =  99.99          DILUTION WS (M/S)       =   99.99
  CAVITY HT (M)         =  5.26         CAVITY HT (M)           =    4.25
  CAVITY LENGTH (M) = 10.76             CAVITY LENGTH (M)       =    6.67
  ALONGWIND DIM (M) = 5.00              ALONGWIND DIM (M)       =   10.00
 CAVITY CONC NOT CALCULATED FOR CRIT WS > 20.0 M/S.  CONC SET = 0.0
     *****************************************
     *** SUMMARY OF SCREEN MODEL RESULTS ***
     *****************************************
  CALCULATION       MAX CONC    DIST TO   TERRAIN
   PROCEDURE        (UG/M**3)   MAX (M)   HT (M)
 --------------   -----------  -------  -------
 SIMPLE TERRAIN     29.17        241.         0.
 ****************************************************
 ** REMEMBER TO INCLUDE BACKGROUND CONCENTRATIONS **
 ****************************************************
```

This approach does not require site-specific meteorological input as calculations are made for a spectrum of possible combinations of wind speed and atmospheric stability (using the Pasquill classes). Concentrations are predicted at specified distances downwind of the source. No consideration of wind direction is required as the output represents the concentrations directly downwind.

Notice that the program prompts the user for the building height, the minimum horizontal building dimension, and the maximum horizontal building dimension in order to consider building downwash problems introduced in Chapter 3. Traditional downwash models included in dispersion models such as SCREEN and indicated on the previous page (for example Huber and Snyder,* Schulman and Hanna†) rely almost exclusively on laboratory data provided by wind tunnel experiments. Physical and scaling constraints limited such experiments to moderate and high wind speeds, and neutral stability, but practical demands to meet regulatory pressures have led to these algorithms being applied more widely.

Of particular concern is the common modelling outcome that critical concentrations in the context of regulatory constraints often occur during stable conditions with light wind. Such-conditions are far removed from those under which the models were developed.

Traffic Pollution in City Streets

In most European cities, poor air quality is often caused by traffic, especially in street canyons with high traffic density. A street canyon is a relatively narrow street between buildings that line up continuously along both sides. The buildings impede the dispersion of the pollutants emitted from traffic, leading to high concentration levels along the street. This in turn causes the high air pollution exposure of pedestrians, car drivers and occupants of buildings immediately adjacent to the roadside. The worse pollution conditions in streets are observed at low wind speeds, where the traffic induced turbulence often is limiting for the pollutant concentrations. Plume models with simple parameterizations of the dispersion in the street, have in many cases been shown to reproduce the observed concentrations well. However, traffic-induced turbulence is not well determined. Equally uncertain is the possible impact of thermal effects inside street canyons on the dispersion conditions. These effects are likely to be mainly of importance in warm climates.

Residence time plays a major role in the chemical reactions that may take place in urban areas. In a single street, the residence time of the air mass can be as short as a few seconds. On the other hand, during anticyclonic pollution episode situations with very low wind speeds, the pollutant residence times within the canopy are at least of the order of several minutes to hours. Stagnation of pollutants can occur in the lowest layers of the urban atmosphere for many hours in the same area.

In recent years the human health effects of particles have attracted increasing attention. The processes governing particle concentrations are complex, especially in urban areas, where re-suspended material in the streets plays a major role. These processes are not well described in present models.

Clearly, the modelling of pollution even within the confines of city streets is subject to many uncertainties, but models are available. Emission factors for different categories of vehicle are usually determined by laboratory experiments. These experiments cannot take into account the variations in the car fleet from one area to another. Emission factors are included in one example of a specialized workbook procedure, the *Design Manual for Roads and Bridges* (DMRB).‡ The procedures described in the manual allow an initial evaluation to be made of the impact on air quality of changes in traffic routes. Simply, the approach involves combining data on vehicle flows along specific

---

* Huber A H and Snyder W H, 1982, 'Wind Tunnel Investigation of the Effects of a Rectangular Shaped Building on Dispersion of Effluents from Short Adjacent Stacks'. *Atm. Env.* 16, 2837–2848.

† Schulman L L and Hanna S R, 1986, 'Evaluation of downwash modifications to the industrial source complex model', *JAPCA*, 36, 258–264.

‡ *Design Manual for Roads & Bridges*, Vol. 11, Sect. 3, Part 1 Air Quality, Department of The Environment, Transport & the Regions, Stationery Office, London, 1999.

sections of roads with emission factors for different types of vehicles of different ages. Middleton* identified several limitations in this simple model. For example, it may use low emission factors for urban traffic and its approach to dispersion beyond 80 m from the road is flawed. However, at such distances this may be unimportant. In addition, DMRB cannot calculate a time series of hourly mean concentrations. Nevertheless it is a familiar modelling tool to traffic planners, and has a role in simple screening. Where an initial study indicates significant changes, a more detailed assessment may be needed. This staged process reflects the general approach to model use introduced in Table 5.5. Appendix 5 indicates availability of a spreadsheet version of DMRB through the Internet.

AEOLIUS

AEOLIUS is the acronym for assessing the environment of locations in urban streets, and is a model developed for local authority use. It calculates hourly concentrations of pollutants within a street canyon at kerbside level. The concentrations calculated are for parallel or along street flow and for across street flow, particularly leeward concentrations where leeward is defined here as the side of the street from where the wind is blowing. Hourly concentrations are calculated for nitrogen oxides, carbon monoxide, sulphur dioxide, benzene 1,3 butadiene, and PM_{10}. Further calculations are made for an hourly concentration and a 98 percentile value of nitrogen oxides as well as a maximum 8 hr concentration of carbon monoxide.
A nomogram version† for modelling traffic pollution in street canyons has been based on the AEOLIUS model. This paper-based method uses tabulated data, graphs and emission factors to provide a screening method for estimating maximum concentrations irrespective of wind direction. The results are similar to those from the computer model. As ever, a screening technique allows a quick estimation of the likely worst pollution problems in road canyons across an urban area.

Two computer versions of AOLEUS are available. A screening version may be run directly from a floppy disk. Input data for the model are summarized in the following box.

AOLIUS Screening version input data	
STREET WIDTH	typical range 10.0 and 30.0 m
STREET HEIGHT	typical range 10.0 and 30.0 m
TRAFFIC SPEED	typical range 5.0 and 15.0 m s^{-1}.
VEHICLE FLOW PER HOUR	typical range 1000 and 4000)
WIND SPEED	typical range 2 and 8 m s^{-1}
POLLUTANT	NOx, CO, PM_{10}, SO_2, C_6H_6 or C_4H_6.
BACKGROUND	the ambient value in the town or city.
YEAR	from 1996 to 2005.
Concentration units of ppb for all pollutants except PM_{10} which is in μg m^{-3}.	

The full version of the model known as AEOLIUSF is a more detailed model than the simpler screening version AEOLIUS. Whereas AEOLIUS asks for runtime screen input as indicated above and fixes some parameters internally to calculate pollution concentrations, AEOLIUSF requires more input data including meteorology. Traffic data for AEOLIUSF must be presented as a one week summary of average flows, with a file for each vehicle type. Vehicle speed is also a one week summary of representative speeds at each hour through the week. Further details of AOLIUS as well as the models themselves are available for download from the Meteorological Office web site.

* Middleton, D R, 1997, *Manual on Modelling: A Guide for Local Authorities*, Met O (APR) Turbulence & Diffusion Note No. 241, The Meteorological Office.

† Buckland A T & Middleton D R, 1999, 'Nomograms for calculating pollution within street canyons', *Atm. Env*, 33, 1017–1036.

Conclusion

Dispersion modelling may be carried out with limited resources, ranging from simple nomograms to the use of public domain software. However, use of such techniques needs an understanding of the underlying principles and limitations of models described in previous chapters. The role of simple models is often to serve as screening techniques to determine whether time and resources need to be invested in more detailed modelling. Such techniques are the topic of Chapter 6.

Exercises

1. A coal-fired furnace in a rural area burns fuel containing 2% sulphur at a rate of 5 tonnes per hour. What is the magnitude and location of the maximum ground level concentration on an overcast day when the wind speeds is 4 m s^{-1}? The effective height of emission is 100 m.

2. An electricity generating station burns residual fuel oil at a rate of 10 tonnes every hour. The oil has a sulphur content of 3% and the final discharge of combustion products into the atmosphere occurs at an effective height of 150 m. On a clear night with wind speeds around 4m s^{-1}, what is the magnitude and location of the maximum ground level concentration contributed by the power station?

3. The world's tallest building at 553 m is the CN Tower in Toronto. Imagine a proposal to build a power plant burning oil as described in Question 2. With wind speeds of 4 m s^{-1} the task is to assess the impact on the tower assuming that the source were located at the worst position:
   a) under day time conditions of strong sunlight; and
   b) under overcast night-time conditions.

4. Using a computer for repetitive calculations, compare the relative impact at ground level over a distance of 15 km for a release from a stack of effective height 300 m and under different conditions of atmospheric stability. Assume a standard release rate and a single wind speed to identify the stability leading to the greatest ground level impact.

5. Compare the impact on ground level concentrations from a source at ground level with one emitting at an effective height of 30 m. Explore the impact over a range of 1 km. Assume that the emission rate is 125 g s^{-1} and the wind speed is 3.5 m s^{-1}.

6. A gas turbine plant is normally fired by natural gas, but has the capacity of running on distillate fuel oil if the gas supply is interrupted. The distillate fuel has a specification sulphur content of 0.3% by mass, and its daily rate of consumption during a gas supply interruption is 484 tonnes. The plant is located in flat terrain at the edge of an urban area. The nearest dwellings are 700 m from the stack, which is 60 m high and has an effective stack diameter of 4.3 m. The efflux velocity during oil firing is 21.5 m s^{-1} and the efflux temperature is 418 K. Assess the significance of using this alternative fuel on the impact at the nearest houses under cloudy conditions with a wind speed at 10 m of 8 m s^{-1} and when the ambient temperature is 10°C.

CHAPTER 7
REFINED COMPUTER MODELS

This chapter aims to introduce some of the so-called refined computer models for atmospheric dispersion. Some of the fundamental differences between the established and 'new generation' models will be outlined, and are significant since they can contribute to differences in predictions. Models are continually being updated and replaced, and so this chapter can give only an indication of the position at the time of writing. So, while model requirements and capability may be summarized here, specifications should be checked against current versions.

Refined Dispersion Models

The first PC versions of mainframe computer models became available from the mid-1980s. Since that time, the growth in the use of personal computers has resulted in a similar increase in the availability of models. Refinements over early mainframe models have included features such as built in error checking as data are entered, on-screen help and options for graphical output. Refined models provide more accurate estimates of a source's impact than simpler screening models and, consequently, require more detailed input data. The model selected for a particular application should be the one that most accurately represents atmospheric transportation, dispersion and chemical transformations in the area under analysis. For example, models have been developed for both simple and complex terrain situations; some are designed for urban applications, while others are designed for rural applications.

More refined modelling of dispersion around stationary sources can be carried out by one of the simple Gaussian plume models. These models predict dispersion patterns of non-reactive pollutants within 50 km of the emission source and are often expected to produce results within a factor of two of the measured values, but this may be optimistic, as Chapter 9 will suggest. Most dispersion models of this type are similar in design and performance, and do not attempt to account for complex situations such as long-range transport and highly reactive chemical emissions.

What distinguishes the various models is their capability to handle different settings. Some of the models are characterized as 'Preferred models' by organizations, such as the US EPA, because they meet certain minimum technical criteria, have undergone field-testing and extensive peer review. This does not make the non-preferred models less suitable for an application, but it provides a documented experience base for the preferred models, which may add credibility to the analysis or eliminate the need for model validation. This will be addressed in Chapter 9. UK regulations and regulators do not specify any specific models.

Important characteristics of some commonly used air quality dispersion models are included in Appendix 3. These models were developed and used mainly in developed countries, but several have been used successfully in developing countries. However, care should always be taken over unusual cases of topography and weather patterns. For example, dispersion models have not been subject to a comparison of model calculations of existing sources with monitored air quality data in tropical weather conditions.

One of the most commonly used models for assessment of pollutant dispersion is from the US EPA. The Industrial Source Complex (ISC) model is used for point (stack), area and volume sources in flat or complex terrain. Two versions exist: the short term version, ISCST for averaging periods of 24 hrs or less, and the long term version, ISCLT for averaging periods of 30 days or longer. However, modelling has been developing in recent years through the release of 'new generation' models. Examples of these new models include the Atmospheric Dispersion Modelling

System (ADMS) in the UK; the AERMOD model in the US and OML from Denmark. They are introduced in this chapter and examples of their comparative use will be given in Chapter 9.

It will be recalled that the atmospheric boundary layer defines the vertical space into which the discharged gases can spread. The characteristics of the boundary layer for both old and new models can be derived from meteorological measurements taken using a standard weather station and converted to represent atmospheric stability using procedures described previously. Most traditional dispersion models, such as the ISC model, are based on the empirical dispersion curves derived from the Prairie Grass experiment in 1956. Table 7.1 summarizes some of the determinants.

Table 7.1 Some Atmospheric Parameters Used in Modelling

Weather conditions	Stability	Monin–Obukhov length (m)	Pasquill–Gifford category
Very sunny summer weather	Extremely unstable	-2	A
Sunny and warm	Moderately unstable	-10	B
Partial cloud during the day	Slightly unstable	-100	C
Overcast day or night	Neutral	Infinity	D
Partial cloud during the night	Slightly stable	100	E
Clear night	Stable	20	F

The newer models calculate the plume spread based on the conditions occurring at the height of the plume. With the older models the plume spread is generally calculated based on ground level conditions. The models also differ in the calculation of dispersion for sunny (unstable) weather conditions. The vertical spread of pollutant is assumed non-Gaussian with the newer models, compared with Gaussian for the older models. This is to take account of the skewed structure of the vertical component of turbulence. Both old and new models use a Gaussian distribution for night-time and cloudy weather conditions.

The ISC model is a steady-state Gaussian plume model used to assess pollutant concentrations from a variety of industrial sources. Like all US EPA models, ISC is available from the EPA website (Figure 7.1) free of charge.

Figure 7.1 The Technology Transfer Network of the US EPA Gives Access to Software Through Its Bulletin Board

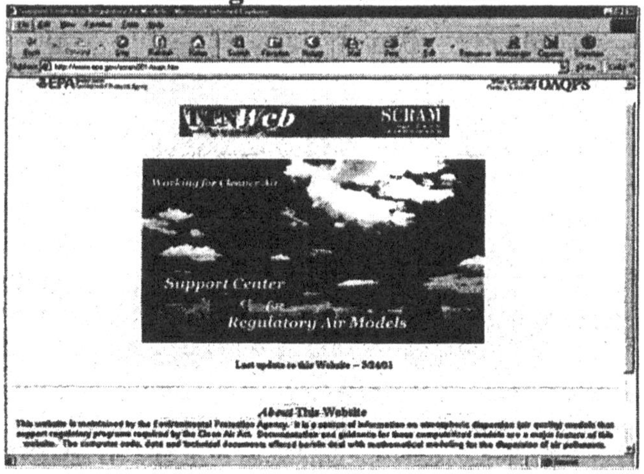

In addition, for a fee, all EPA models are available from the US National Technical Information Service (NTIS), Springfield, VA 22161, US.

The ISC model accounts for settling and dry deposition of particulate matter; building obstruction wake effects; plume rise as a function of downwind distance; multiple, but separate, point area and volume sources. There is also an ability to analyse concentrations in any type of terrain. It can estimate from hourly to annual pollutant concentrations. Throughout the world the application of ISC is extensive, and it has been recommended and used widely for both urban and rural areas. It is similar to models developed by larger manufacturing companies for internal use, while many commercial organizations have put the ISC code into a Windows format. The enhancements include user-friendly data input, the capability of easily plotting the output, custom output summaries, and technical support. Some of these companies offer training in the use of models, both in the US and overseas. One commercial version from Trinity Consultants Inc under the BREEZE name will be used as an illustration, although versions are also available from other suppliers such as Lakes Environmental Software, Bee-Line Software, etc. Web addresses for such companies are included in Appendix 5.

Models such as ISC are more sophisticated in their structure and capabilities than the simple screening models introduced previously, but they can be applied in a basic manner where the necessary data are difficult to obtain or in order to determine the value of collecting further data. For example, in a complex airshed with several sources and differing meteorological conditions, it may be appropriate to calculate the impacts of each major source individually using simplified meteorology for that source. The impacts from all of the sources can then be added together, producing estimates that are not as precise as a multi-source model but which may give a reasonable indication of the overall impact as a basis for local air quality management.

In complex situations, significant effort and professional judgement is often necessary for developing an emissions inventory, including defining relevant sources, in collecting local meteorological data and in selecting combinations of conditions to be modelled. For example, in an industrial area, it might be appropriate to group the sources into different types, such as major sources, smaller industrial or municipal sources, and indeterminate residential emissions. Identification of sources and estimation of the emissions inventory for use in the model would be a significant task. For major sources, site specific details should be obtained. Smaller industrial sources might be handled by aggregating into groups or by defining typical characteristics. Residential sources would have to be partitioned in accordance with some estimates of population density or housing type. Efforts would have to be made to understand the local meteorology in some detail, addressing variability across the area, patterns of inversions and perhaps day/night variations to reflect patterns of residential emissions. An example in the final chapter illustrates some of these problems. Credible modelling in such complex situations requires significant effort and resources. Collection and interpretation of the data required as input can represent 50% or more of the overall work.

The Industrial Source Complex Model (ISC)

The ISC short term model (ISCST) provides options to model emissions from a wide range of sources that might be present at a typical industrial site. Emission sources are identified as fitting into four basic types of sources, ie, point sources, volume sources, area sources and open pit sources. The basis of the model is the point source, steady-state Gaussian plume equation, which is used with some modifications to model simple point source emissions from stacks, emissions from stacks that experience the effects of aerodynamic downwash due to nearby buildings, isolated vents, multiple vents, storage piles, conveyor belts, etc. It will be noted that such emissions include particulate matter, of which little has been said so far.

Dry deposition of particulate matter assumes that particles are brought to the surface by a combination of turbulent diffusion and gravitational settling. Once near the surface, they may be removed from the atmosphere and deposited on the surface. This removal is modelled by assuming that the deposition flux of material to the surface is equal to the product of the airborne concentration just above the surface and the deposition velocity. As a plume containing particles is transported downwind, this deposition near the surface reduces the concentration of particles in the plume, and so alters the vertical distribution of the remaining particles. Moreover, larger particles move steadily nearer the surface at a rate equal to their gravitational settling velocity. The effect is to reduce the height of the plume centre line, and the concentration distribution in the vertical plane is no longer Gaussian.

In models such as ISC, a corrected source-depletion model is used to obtain a 'vertical term' that incorporates both the gravitational settling of the plume and the removal of plume mass at the surface. These effects are incorporated as modifications to the Gaussian plume equation. For large travel-times, the tilted plume approximation is not valid. Therefore, a minimum value of zero is imposed. In effect, this limits the settling of the plume centre line, although the deposition velocity continues to account for gravitational settling near the surface. The size and density of the particles being modelled influence the total deposition velocity. To account for this, models such as ISC allow multiple particle-size categories to be defined for a specific source. The user provides the mass-mean particle diameter (μm), the particle density (g cm^{-3}), and the mass fraction for each category being modelled.

Hourly meteorological data records define the conditions for plume rise, transportation, diffusion and deposition. The model estimates the concentration or deposition value for each source and receptor combination for each hour of input meteorology, and calculates user-selected short term averages. For deposition values, either the dry deposition flux, the wet deposition flux or the total deposition flux may be estimated. The total deposition flux is simply the sum of the dry and wet deposition fluxes at a particular receptor location. There is an option to select averages for the entire period of input meteorology.

The ISC long term model (ISCLT) provides options to model emissions from the same types of sources as provided by the short term model, and similarly is based on the straight-line, steady-state Gaussian plume equation. The ISC long term model uses input meteorological data that have been summarized into joint frequencies of occurrence for particular wind speed classes, wind direction sectors and stability categories. These summaries, called STAR summaries for STability ARray, may include frequency distributions over a monthly, seasonal or annual basis. The long term model can be used to calculate concentration or dry deposition values for each separate STAR summary input and/or for the combined period covered by all available STAR summaries.

The two basic types of inputs that are needed to run the ISC models are an input run-stream file and a meteorological data file. The run-stream setup file contains the selected modelling options, as well as source location and parameter data, receptor locations, meteorological data file specifications, and output options. Many of these elements are evident in Figure 7.2, which shows a window from a commercial version of ISC.

Figure 7.2 Setting the Modelling Options in BREEZE: a Commercial Version of ISC in Windows Format

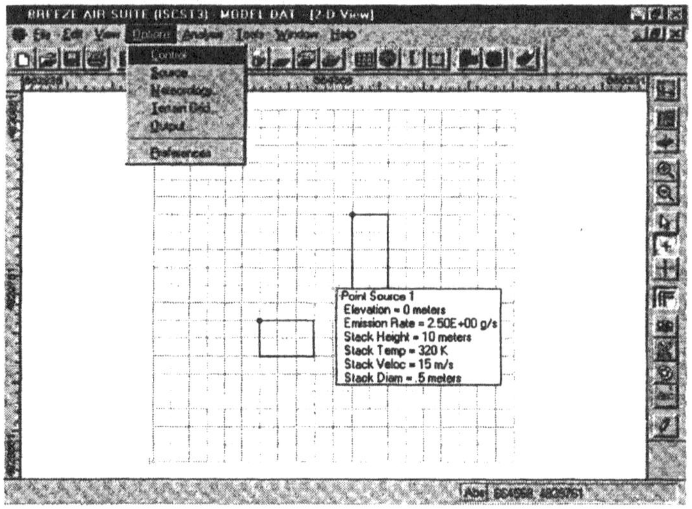

Since the ISC models were designed to support US statutory modelling requirements, the regulatory modelling options are the default mode of operation for the models. These options are evident from a printout after a model run as in the following information box.

```
*** ISCST – VERSION 93109 ***    *** Model run for ISCST Atmospheric Dispersion Calculation    ***    25/08/01
***      21:07:56
PAGE  1
*** MODELING OPTIONS USED:  CONC  URBAN  ELEV        DFAULT
***    MODEL SETUP OPTIONS SUMMARY      ***
- - - - - - - - - - - - - - - - - - - - - - - - - - - - - - - - - - - - - - - - - - - - - - - - - - - - -
**Model Is Setup For Calculation of Average CONCentration Values.
**Model Uses URBAN Dispersion.
**Model Uses Regulatory DEFAULT Options:
1. Final Plume Rise.
2. Stack-tip Downwash.
3. Buoyancy-induced Dispersion.
4. Use Calms Processing Routine.
5. Not Use Missing Data Processing Routine.
6. Default Wind Profile Exponents.
7. Default Vertical Potential Temperature Gradients.
8. "Upper Bound" Values for Supersquat Buildings.
9. Half-life of 4 hrs for URBAN SO2.
**Model Accepts Receptors on ELEV Terrain.
**Model Assumes No FLAGPOLE Receptor Heights.
**Model Calculates 1 Short Term Average(s) of:  1-HR
**This Run Includes: 1 Source(s);  1 Source Group(s); and   441 Receptor(s)
**The Model Assumes A Pollutant Type of: SO2
**Model Set To Continue RUNning After the Setup Testing.
**Output Options Selected:
Model Outputs Tables of Highest Short Term Values by Receptor (RECTABLE Keyword)
**NOTE:  The Following Flags May Appear Following CONC Values: c for Calm Hours
m for Missing Hours
b for Both Calm and Missing Hours
**Misc. Inputs:  Anem. Hgt. (m) =   10.00 ;   Decay Coef. =  0.4810E-04 ;   Rot. Angle =    0.0
Emission Units = GRAMS/SEC   ; Emission Rate Unit Factor =  0.10000E+07
Output Units   = MICROGRAMS/M**3
**Input Runstream File:          H:\MODELS\ISCSTEX\Example.DAT
**Output Print File:             H:\MODELS\ISCSTEX\Example.LST
```

The user may select either rural or urban dispersion parameters, depending on the characteristics of the source location. There is also an option of calculating concentration values or deposition values for a particular run.

The model can handle multiple sources, including point, volume, area and open pit source types. Line sources may also be modelled as a string of volume sources or as elongated area sources. Several source groups may be specified in a single run, with the source contributions combined for each group. This is particularly useful for applications where combined impacts may be of concern, while the combined impacts from all background sources (and the source under investigation) are needed to demonstrate compliance with ambient air quality standards. Source emission rates can be treated as constant throughout the modelling period, or may be varied by month, season, hour-of-day, or other optional periods of variation.

The example entries in the BREEZE version of ISC indicated in Figure 7.2 were made after selecting the Parameters tab within Source options. The pollutant of interest here is sulphur dioxide, and the 1 hr averaging period was selected as was evident from the previous printout.

The ISC models allow the user to specify multiple receptor networks in a single run, and may also mix Cartesian grid receptor networks and polar grid receptor networks in the same run. This is useful for applications where a coarse grid may be needed over the whole modelling domain, with a denser grid in the area of maximum expected impacts. There is also flexibility in specifying the location of the origin for polar receptors, other than the default origin at (0,0) in x,y co-ordinates. Elevated receptor heights may be examined in order to model the effects of terrain above (or below) stack base, and may specify receptor elevations above ground level to model flagpole receptors. Factors such as these are shown in the printout represented in the following information box.

*** MODELING OPTIONS USED: CONC URBAN ELEV DFAULT

*** NETWORK ID: Example ; NETWORK TYPE: GRIDCART ***

* ELEVATION HEIGHTS IN METERS *

Y-COORD (METERS)	X-COORD (METERS) -1000.00	-900.00	-800.00	-700.00	600.00	-500.00	400.00	-300.00	-200.00
1000.00	0.00	0.00	0.00	0.00	0.00	0.00	0.00	0.00	0.00
900.00	0.00	0.00	0.00	0.00	0.00	0.00	0.00	0.00	0.00
800.00	0.00	0.00	0.00	0.00	0.00	0.00	0.00	0.00	0.00
700.00	0.00	0.00	0.00	0.00	0.00	0.00	0.00	0.00	0.00
600.00	0.00	0.00	0.00	0.00	0.00	0.00	0.00	0.00	0.00

etc

Following through the example started above in the BREEZE version, the Cartesian Grid tool was selected from the Toolbar in Figure 7.2. The mouse pointer was placed at the SW (bottom left) corner at about -1000, -1000 in the area where the grid was to be drawn. Dragging upwards and to the right drew a receptor grid as indicated in the Figure. A Gridded Receptor Network dialogue box appeared allowing the coordinates to be set. The output from the example run is illustrated below.

*** NETWORK ID: example ; NETWORK TYPE: GRIDCART ***

*** X-COORDINATES OF GRID ***
(METERS)

-1000.0, -900.0, -800.0, -700.0, -600.0, -500.0, -400.0, -300.0, -200.0, -100.0,
0.0, 100.0, 200.0, 300.0, 400.0, 500.0, 600.0, 700.0, 800.0, 900.0, 1000.0,

*** Y-COORDINATES OF GRID ***
(METERS)

-1000.0, -900.0, -800.0, -700.0, -600.0, -500.0, -400.0, -300.0, -200.0, -100.0,
0.0, 100.0, 200.0, 300.0, 400.0, 500.0, 600.0, 700.0, 800.0, 900.0, 1000.0,

Under Meteorology Options, data from Birmingham Airport (UK) were selected; the same data that generated the wind rose in Chapter 3. The model printout is shown in the following box.

METEOROLOGICAL DATA PROCESSED BETWEEN START DATE: 90 1 1 1 AND END DATE: 90 1 10 24
NOTE: METEOROLOGICAL DATA PROCESSED WILL DEPEND ON WHAT IS INCLUDED IN THE DATA FILE.

*** UPPER BOUND OF FIRST THROUGH FIFTH WIND SPEED CATEGORIES ***
(METERS/SEC)
1.54, 3.09, 5.14, 8.23, 10.80,

*** WIND PROFILE EXPONENTS ***

STABILITY CATEGORY	WIND SPEED CATEGORY 1	2	3	4	5	6
A	.15000E+00	.15000E+00	.15000E+00	.15000E+00	.15000E+00	.15000E+00
B	.15000E+00	.15000E+00	.15000E+00	.15000E+00	.15000E+00	.15000E+00
C	.20000E+00	.20000E+00	.20000E+00	.20000E+00	.20000E+00	.20000E+00
D	.25000E+00	.25000E+00	.25000E+00	.25000E+00	.25000E+00	.25000E+00
E	.30000E+00	.30000E+00	.30000E+00	.30000E+00	.30000E+00	.30000E+00
F	.30000E+00	.30000E+00	.30000E+00	.30000E+00	.30000E+00	.30000E+00

*** VERTICAL POTENTIAL TEMPERATURE GRADIENTS ***
(DEGREES KELVIN PER METER)

STABILITY CATEGORY	WIND SPEED CATEGORY 1	2	3	4	5	6
A	00000E+00	00000E+00	00000E+00	00000E+00	00000E+00	00000E+00
B	00000E+00	00000E+00	00000E+00	00000E+00	00000E+00	00000E+00
C	00000E+00	00000E+00	00000E+00	00000E+00	00000E+00	00000E+00
D	00000E+00	00000E+00	00000E+00	00000E+00	00000E+00	00000E+00
E	.20000E-01	.20000E-01	.20000E-01	.20000E-01	.20000E-01	.20000E-01
F	.35000E-01	.35000E-01	.35000E-01	.35000E-01	.35000E-01	.35000E-01

*** THE FIRST 24 HRS OF METEOROLOGICAL DATA ***

FILE: H:\MODELS\BIRBIR90.BIN FORMAT: UNFORM
SURFACE STATION NO.: 3534 UPPER AIR STATION NO.: 3534
NAME: BIRMINGHAM NAME: BIRMINGHAM
YEAR: 1990 YEAR: 1990

YEAR	MON	DAY	HR	FLOW VECTOR	SPEED (M/S)	TEMP (K)	STAB CLASS	MIXING HT (M) RURAL	URBAN
90	1	1	1	331.0	4.12	276.5	4	732	1149
90	1	1	2	348.0	5.66	276.5	4	1034.8	1606.8
↓	↓	↓	↓	↓	↓	↓	↓	↓	↓
90	1	1	24	50,0	4.12	278.1	4	724.2	1141.2

*** NOTES: STABILITY CLASS 1=A, 2=B, 3=C, 4=D, 5=E AND 6=F.
FLOW VECTOR IS DIRECTION TOWARD WHICH WIND IS BLOWING.

The short term model includes dry and wet deposition algorithms. The former requires additional meteorological input variables, such as Monin–Obukhov length and surface friction velocity that are provided by the PCRAMMET pre-processor. The wet deposition algorithm in the short term model also needs precipitation data, which are options in PCRAMMET pre-processed data. The long term model uses STAR (STability ARray) summaries described earlier. These STAR summaries may be generated from sequential data files using the STAR utility program available on the SCRAM Bulletin Board System (Figure 7.1) or by the MPRM meteorological processor for on-site data.

The basic types of printed output available with the short term model are:

- Summaries of high values (highest, second highest, etc) by receptor for each averaging period and source group combination.
- Summaries of overall maximum values (e.g., the maximum 50) for each averaging period and source group combination.
- Tables of concurrent values summarized by receptor for each averaging period and source group combination for each day of data processed.

In the example being worked here, under Output Options the first highest option was selected as identified in the printout earlier. The output is illustrated in the following box. This specifies the hour of the meteorological data used in the calculation (for example 90010705).

```
          *** ISCST - VERSION 93109 *** Model run for ISCST Atmospheric Dispersion Calculation
***25/08/01***21:07:56
PAGE  7
*** MODELING OPTIONS USED:  CONC  URBAN  ELEV       DFAULT
*** THE 1ST HIGHEST 1-HR AVERAGE CONCENTRATION VALUES FOR SOURCE GROUP ALL
***INCLUDING SOURCE(S):     STACK1  ,
                    *** NETWORK ID: Example;  NETWORK TYPE: GRIDCART ***
                    ** CONC OF SO2      IN MICROGRAMS/M**3                          **
Y-COORD                                     X-COORD (METERS)
(METERS)    -1000.00       -900.00        -800.00        -700.00        -600.00
1000.0      37.87779       50.83891       82.92991       86.27933       50.65566
            (90010707)     (90010705)     (90010705)     (90010705)     (90010705)
900.0       68.91889       44.63727       63.69110       100.07146      89.28758
            (90010707)     (90010707)     (90010705)     (90010705)     (90010705)
800.0       86.98759       83.11481       53.31514       81.32254       119.59391     etc.
            (90010707)     (90010707)     (90010707)     (90010705)     (90010705)
*** THE SUMMARY OF HIGHEST  1-HR RESULTS ***
** CONC OF SO2      IN MICROGRAMS/M**3                **
GROUP ID   AVERAGE CONC   (YYMMDDHH)    RECEPTOR  (XR, YR, ZELEV, ZFLAG) OF TYPE
GRID-ID
--------------------------------------------------------------------------------------
---
ALL    HIGH  1ST HIGH VALUE IS  546.78741  ON 90010617:  AT (  100.00,     0.00,     0.00,   0.00) GC
Example
*** RECEPTOR TYPES:  GC = GRIDCART
                 GP = GRIDPOLR
                 DC = DISCCART
                 DP = DISCPOLR
                 BD = BOUNDARY
*** Message Summary For ISC2 Model Execution ***
--------- Summary of Total Messages --------
A Total of        0 Fatal Error Message(s)
A Total of        0 Warning Message(s)
A Total of        4 Informational Message(s)
A Total of        4 Calm Hrs Identified
  ******** FATAL ERROR MESSAGES ********
        ***  NONE  ***
  ********  WARNING MESSAGES  ********
        ***  NONE  ***
  ************************************
  *** ISCST Finishes Successfully ***
  ************************************
```

For the long term model, output tables of values for each receptor, and/or tables of overall maximum values may be produced. In addition to the tabular printed outputs, the ISC models provide options for several types of file output products. One of these options for ISCST is to output an unformatted ('binary') file of all concentration and/or deposition values as they are calculated. These files are often used for subsequent processing of the data.

As well as unformatted concentration files, ISCST provides options for three additional types of file outputs. One option is to generate an ASCII formatted file with the same results that are included in the unformatted post-processing file. Another option is to generate a file of (x,y) coordinates and design values (for example the second highest values at each receptor for a particular averaging period and source group combination) that can be easily imported into many graphics plotting packages to generate isopleths of the concentration and/or deposition values. A third option of the ISCST model is to generate a file of all occurrences when a concentration or deposition value equals or exceeds a threshold specified by the user. Reference will be made later to commercial packages that incorporate software to improve visualization of the model output.

AERMOD

AERMOD was designed for applications similar to those handled by the ISC model. In 1991, the Regulatory Model Improvement Committee (AERMIC) was set up by the EPA and others to accelerate the inclusion of advances in atmospheric boundary layer understanding into the EPA regulatory models. The primary focus of AERMIC was to improve the scientific basis of regulatory dispersion models by introducing increased knowledge of turbulence and diffusion in the planetary boundary layer that had been gained over the previous two decades. In 1994, AERMIC completed the first version of the AERMIC Model (AERMOD) and its meteorological pre-processor (AERMET). AERMOD is a steady-state plume-based model designed for estimating near-field impacts from a variety of industrial source types. In developing AERMOD, certain design criteria were followed to produce a model with the desired regulatory attributes, namely that the model should be:

- robust in estimating regulatory design concentrations (ie provide reasonable concentration estimates over a wide range of conditions with minimal discontinuities);
- easily implemented with reasonable input requirements and computer resource needs;
- based on up-to-date science that captures the essential physical processes while remaining fundamentally simple; and
- easily revised as the science evolves.

AERMIC adopted the modular ISC architecture as the framework for AERMOD as this facilitated the incorporation of new and improved algorithms. AERMOD consists of a meteorological interface that computes vertical profiles of wind, turbulence and temperature. There are dispersion algorithms for the daytime convective boundary layer; dispersion algorithms for the night-time stable boundary layer. Surface, near-surface and elevated releases (buoyant and non-buoyant) are treated, as is complex terrain.

The terrain pre-processor for AERMOD is AERMAP, which uses gridded terrain data (for example US Geological Survey Digital Elevation Mapping data) to model terrain influence on the atmospheric flow. A major part of the developmental evaluation of AERMOD included comparisons of model estimates with measurements from a variety of field studies. The purpose of the evaluation was primarily diagnostic, ie, to identify and correct problems in the model during its development. AERMOD was compared with standard data sets including the Prairie Grass passive surface release (rural) data, Kincaid elevated, buoyant release (rural) data, and Indianapolis elevated, buoyant release (urban) data. In summary, for surface releases, AERMOD had a slight tendency to under-predict the highest concentrations in both convective and stable conditions, for both urban and rural elevated releases. It also exhibited insignificant bias in convective conditions and again only a slight tendency toward under-prediction in stable conditions.

The output from AERMOD resembles that from ISC as the input and output examples in the following boxes indicate. The scenario involved a single point source of sulphur dioxide from a 60 m stack of diameter 2.1 m with a buoyant release at a temperature of 443 K and an efflux velocity of 13.5 m.s^{-1}. The scenario involved a rural setting with relatively flat terrain surrounding the plant and in a cartesian receptor network.

```
*** AERMOD - VERSION 99211 ***   *** An Example from the AERMOD Model                  ***    12/22/00
                                 *** single source example                             ***    10:39:31
                                                                                              PAGE  1
*** MODELING OPTIONS USED:          CONC          ELEV
                          ***   MODEL SETUP OPTIONS SUMMARY      ***
- - - - - - - - - - - - - - - - - - - - - - - - - - - - - - - - - - - - - - - - - - - - - - - - - - -
 **Model Is Setup For Calculation of Average CONCentration Values.
 **Model Uses RURAL Dispersion Only.
 **Model Uses User-Specified Options:
       1. Stack-tip Downwash.
       2. Model Accounts for ELEVated Terrain Effects.
 **Model Assumes No FLAGPOLE Receptor Heights.
 **Model Calculates  1 Short Term Average(s) of:  24-HR
 **This Run Includes:     1 Source(s);      1 Source Group(s); and     121 Receptor(s)
 **The Model Assumes A Pollutant Type of:  SO2
 **Model Set To Continue RUNning After the Setup Testing.
 **Output Options Selected:
        Model Outputs Tables of Highest Short Term Values by Receptor (RECTABLE Keyword)
**NOTE: The Following Flags May Appear Following CONC Values: c for Calm Hours
                                                              m for Missing Hours
                                                              b for Both Calm and Missing Hours
**Misc. Inputs: Base Elev. for Pot. Temp. Profile (m MSL) = 0.00; Decay Coef = 0.0000; Rot. Angle = 0.0
                Emission Units = GRAMS/SEC            ; Emission Rate Unit Factor =   0.10000E+07
                Output Units   = MICROGRAMS/M**3
**Input Runstream File: C:\DATA\SO2.DAT      ; **Output Print File: C:\DATA\SO2.LST
```

A module in AERMET computes mixing heights in the absence of appropriate upper air sounding data, such as when upper air sounding data are unavailable, or when local times, which correspond to radiosonde ascents (typically 0000 and 1200 GMT), do not occur near sunrise. The method estimates the surface similarity parameters of friction velocity, sensible heat flux, temperature scale, and Monin–Obukhov length using routinely collected meteorological variables of cloud cover, ceiling height, wind speed, and temperature, and estimates of surface roughness. These computed parameters are seen in the following extract of a list file together with a term not used before, the Bowen Ratio. This is a measure of the amount of moisture available at the Earth's surface.

```
1 *** AERMOD - VERSION 99211 ***   *** An Example from the AERMOD Model                 ***    12/22/00
                                   *** single source example                            ***    10:39:31
                                                                                               PAGE 11
 *** MODELING OPTIONS USED:  CONC          ELEV
              *** UP TO THE FIRST 24 HRS OF METEOROLOGICAL DATA ***
     Surface file:   C:\DATA\AMT1.SFC
     Profile file:   C:\DATA\AMT1.PFL
Surface format:
(3(I2,1X),I3,1X,I2,1X,F6.1,1X,3(F6.3,1X),2(F5.0,1X),F8.1,1X,F6.3,1X,2(F6.2,1X),F7.2,1X,F5.0,3(1X,F6.1))
Profile format: (4(I2,1X),F6.1,1X,I1,1X,F5.0,1X,F7.2,1X,F7.1,1X,F6.1,1X,F7.2)
Surface station no.:    3534                  Upper air station no.:     0
                  Name: BIRMINGHAM                                  Name: UNKNOWN
                  Year:     99                                      Year:     99
First 24 hrs of scalar data
YR MO DA JDA HR H0 U*  W*  DT/DZ ZICNV ZIMCH M-O LEN Z0 BOWEN ALBEDO REF WS WD HT REF TA HT
- - - - - - - - - - - - - - - - - - - - - - - - - - - - - - - - - - - - - - - - - - - - - - - -
 99 01 01   1    1 -64.0 1.111 -9.000 -9.000  -999.    2182.  1937.1    1.00 1.62     1.00     6.20     141.  9.1 280.9   2.0
 99 01 01   1    2 -64.0 1.111 -9.000 -9.000  -999.    2182.  1937.1    1.00 1.62     1.00     6.20     138.  9.1 280.9   2.0
.
.
 99 01 01  1    23 -64.0 1.588 -9.000 -9.000  -999.    3278.  5655.7    1.00 1.62     1.00     8.80     170.  9.1 280.9   2.0
99 01 01   1    24 -64.0 1.387 -9.000 -9.000  -999.    2875.  3768.     1.00 1.62     1.00     7.70     160.  9.1 280.9   2.0
```

The results appear for each coordinate location on the cartesian network, with the relevant date as indicated below.

```
*** THE  2ND HIGHEST 24-HR AVERAGE CONCENTRATION  VALUES FOR SOURCE GROUP: ALL   ***
                    INCLUDING SOURCE(S):   SRC1  ,
            *** NETWORK ID: GRD1   ; NETWORK TYPE: GRIDCART ***
              ** CONC OF SO2    IN MICROGRAMS/M**3                  **
                                        X-COORD (METERS)
Y-COORD   -1500.00             -1200.00             -900.00              -600.00
(METERS)
1500.0    1.36303c(99102924)   1.68035 (99091824)   2.03404 (99092224)   2.12873 (99010124)
1200.0    1.47345 (99042024)   1.72083c(99102924)   2.21433 (99102324)   2.41872 (99091824)
900.0     1.63844 (99091924)   1.84021m(99031124)   2.42023 (99091824)   3.05959 (99092224)
600.0     2.28230 (99091924)   2.67972 (99091924)   3.03674m(99040124)   3.79113 (99091824)
300.0     2.16495m(99082424)   2.56093m(99040124)   3.88171 (99091924)   5.11453 (99091924)
0.0       2.86691 (99101524)   3.19126 (99101524)   3.39145 (99101524)   4.05681m(99070924)
                                                                                      etc
```

The list file concludes by reporting the key characteristic, in this case the second highest 24 hr value.

```
1 *** AERMOD - VERSION 99211 ***   *** An Example from the AERMOD Model              ***     12/22/00
                                   *** single source example                         ***     10:39:31
                                                                                             PAGE  15
 *** MODELING OPTIONS USED:  CONC           ELEV
                         *** THE SUMMARY OF HIGHEST 24-HR RESULTS ***
                              ** CONC OF SO2    IN MICROGRAMS/M**3                           **
                                 DATE                         NETWORK
GROUP ID                 AVERAGE CONC (YYMMDDHH) RECEPTOR (XR, YR, ZELEV, ZFLAG) OF TYPE GRID-ID
------------------------------------------------------------------------------------------------
ALL HIGH 2ND HIGH VALUE IS 6.57062m ON 99011724:           AT (300.00, 600.00, 0.00, 0.00)    GC     GRD1

 *** RECEPTOR TYPES:  GC = GRIDCART
                      GP = GRIDPOLR
                      DC = DISCCART
                      DP = DISCPOLR
```

A message summary is also generated, as shown below.

```
1 *** AERMOD - VERSION 99211 ***   *** An Example from the AERMOD Model              ***     12/22/00
                                   *** single source example                         ***     10:39:31
                                                                                             PAGE  16
 *** MODELING OPTIONS USED:  CONC           ELEV
                              *** Message Summary For AERMOD Execution ***
 ------ Summary of Total Messages ------
A Total of       0 Fatal Error Message(s)
A Total of     394 Warning Message(s)
A Total of    2023 Informational Message(s)
A Total of     333 Calm Hrs Identified
A Total of    1690 Missing Hrs Identified ( 19.29 Percent)
 CAUTION!:  Number of Missing Hours Exceeds 10 Percent of Total!
            Data May Not Be Acceptable for Regulatory Applications.
            See Section 5.3.2 of "On-site Meteorological Program Guidance
            for Regulatory Modeling Applications" (EPA-450/4-87-013)
   ******** FATAL ERROR MESSAGES ********
         ***  NONE  ***
********  WARNING MESSAGES  ********
RE W214   31 RECART: ELEV Input Inconsistent With Option: Defaults Used   GRD1
MX W432    23 METQA : Friction Velocity Out-of-Range.  KURDAT=     99010123
...
MX W432  8607 METQA : Friction Velocity Out-of-Range.  KURDAT=     99122515
MX W432  8608 METQA : Friction Velocity Out-of-Range.  KURDAT=     99122516

   ************************************
   *** AERMOD Finishes Successfully ***
   ************************************
```

Windows versions of AERMOD are available commercially.

ADMS

Like AERMOD, ADMS is an example from the UK of the latest generation PC-based models for the atmospheric dispersion of passive, buoyant or slightly dense releases from single or multiple sources. The fundamental difference from older models such as R-91 and ISC is that ADMS, like AERMOD, uses an up-to-date parameterization of the boundary layer as suggested in Chapter 3. By defining the boundary structure in terms of measurable physical parameters, a realistic representation of the changing dispersion with height is made possible. The result is claimed to offer a more accurate and soundly based prediction of the concentration of pollutants than the older models that rely on imprecisely defined Pasquill–Gifford Stability Categories. Features of ADMS include the following:

- Concentration distributions are Gaussian in stable and neutral conditions, but the vertical distribution is non-Gaussian in convective conditions to take account of the skewed structure of the vertical component of the turbulence.
- Plume spread depends on the local wind speed and turbulence and so depends on plume height. This contrasts with the Pasquill–Gifford approach in which plume spread is independent of height.
- A meteorological pre-processor calculates the boundary layer parameters from a variety of input data: for example wind speed, day, time, cloud cover or wind speed, surface heat flux and boundary layer height. Meteorological data may be statistically analysed or raw, hourly averaged data.
- Calculation of averages of mean concentration, deposition and radioactivity, and mean concentration percentiles for averaging times ranging from seconds to a year. For shorter averaging times (1 hr or less) estimates of fluctuations in the concentration, including peaks, are included. Calculated concentrations may be compared to air quality limits and guidelines such as the UK National Air Quality Strategy by selecting the appropriate averaging time and percentile to be calculated.
- For short emissions, time-integrated results allow estimation of dose.
- Modules allow for the effects of plume rise, complex terrain, buildings, coastlines and the calculation of concentration fluctuations and radioactive decay.

As with all models, there are limitations. These include allowance for calm wind conditions and rapidly changing weather conditions and the fact that the model is time independent so no time steps can be included. The example in Chapter 9 will illustrate the significance of this. The input requirements of ADMS are as follows:

- Emissions – source location, emission rate, also volume flow rate, temperature of stack emissions and height and diameter of stacks.
- Meteorology – basic data from one site (for example wind speed, wind direction, temperature, cloud cover) or derived quantities (for example Monin–Obukhov length, boundary layer height). The developers of ADMS supply a number of sample meteorological files suitable for simple test calculations or to examine the effects of a range of individual stability categories on dispersion. If there is a meteorological station on site, the data collected from it can normally be made into a format suitable for use in ADMS. Otherwise, data collected at a large number of UK meteorological stations and world-wide stations are available from the UK Meteorological office, in a format suitable for the model. The model pre-processes raw data using a selection process that is based on the number of variables available.
- Topography – complex terrain has a significant effect on the location and value of the maximum surface concentration where there are slopes, typically greater than 1:10. ADMS uses an airflow model, FLOWSTAR, and calculates the effect of the terrain on the air flow and hence on dispersion. Allowance is made for changes in surface elevation and roughness, and for the stability of the atmosphere. Receptor coordinates, elevations, height above ground and height

scales are produced by the AERMAP terrain preprocessor for input to AERMOD. AERMAP uses Digital Elevation Model (DEM) terrain data produced by the US Geological Survey (USGS), or other equivalent data. Terrain elevation data may also be loaded automatically from OS Landform Panorama digitised terrain data (1:50,000).

The influence of buildings on dispersion was introduced in Chapter 3. The ADMS building module was developed from work on an isolated building or a few buildings that are grouped close together on a site with the stack above a main building. Validation work has shown these arrangements to be treated well. What is not well represented by the model is a very complex site with buildings that can be very far apart. This is because the model reduces all the buildings into a single equivalent wind aligned building at the height of the building specified by the user as the main building in the ADMS interface. This main building is not necessarily the tallest building but is the building that will have the most effect on dispersion. If there is a row of identical buildings on a site, for example a row of warehouses or a group of storage tanks, they can be grouped together and modelled as one larger building to reduce the run time significantly. The model ignores buildings with a height less than one-third of the stack height. However, this is a rule of thumb which although found to be useful will not always be true for all configurations of sources and buildings in the real environment.

Like AERMOD, concentration predictions have been validated against all standard data sets (Prairie Grass, Kincaid, Indianapolis etc). Additional comparisons have been made with LIDAR data and wind tunnel data.*

OML

A third example of the latest generation of model is OML, an acronym for the Danish term 'Operationelle Meteorologiske Luftkvalitetsmodeller'. Developed by the National Environmental Research Institute, Roskilde, Denmark, the OML model is a modern Gaussian plume model, based on boundary layer scaling instead of relying on Pasquill stability classifications. It belongs to the same class of models as ADMS and is intended for use over distances up to about 20 km from a source that is typically one or more stacks, and possibly area sources. The OML model is the recommended model to be used for environmental impact assessments when new industrial sources are planned in Denmark. It can be used for both high and low sources, but is not suitable for complex terrain conditions. The model requires information on emission and meteorology on an hourly basis. It computes a time series of concentrations at user-specified receptor points, from which statistics are extracted and presented to the user.

In the OML model, the Gaussian dispersion parameters σ_y and σ_z are not – as in traditional operational models – functions only of stability category and distance from the source. Instead, they are continuous functions of several boundary layer parameters. Hence, the model does not use traditional discrete stability categories, but describes dispersion processes in terms of basic boundary-layer scaling parameters, such as friction velocity, Monin–Obukhov length, and the convective velocity scale. Thus, in common with similar models of the latest generation described earlier in this chapter, a pre-processor must process meteorological measurements before being used by the model. The pre-processor typically takes input as hourly meteorological measurements from a synoptic or analogous surface station, and twice-daily vertical profiles of temperature from a nearby radiosonde station. Output is in this case hourly values of turbulence parameters: most essentially sensible heat flux, Monin–Obukhov length, friction velocity and mixing height. More specialized versions of the pre-processor take non-standard input such as mast measurements instead of synoptic surface data. For use in Denmark, processed meteorological data are available off the shelf for many locations. The main elements of the meteorological pre-processor have been documented.†* The basis of plume rise modelling is that proposed by Briggs. In contrast to most

* Appendix C of *Validation of the UK-ADMS Dispersion Model and Assessment of its Performance Relative to R-91 and ISC using Archived Lidar Data*, 1995, study commissioned by Her Majesty's Inspectorate of Pollution, published by the Environment Agency.

† Berkowicz, R and Prahm, L P, 1982, 'Sensible heat flux estimated from routine meteorological data by the resistance method', *J. App. Met.*, 21, 1845–1864.

conventional models, penetration of the plume into the atmosphere above the mixing layer is not simulated as an on/off process. Instead, the extent of plume penetration is considered.

The model is Gaussian. So, there is an inherent assumption that the plume has a straight centre line, pointing along the assigned wind direction. The concentration distribution is Gaussian, both horizontally and vertically. Under low wind conditions, a Gaussian model does not perform well, because the basic assumptions underlying the model are violated. The model calculates hourly averaged concentration values.

In its current version, the model takes no account of chemical reactions or deposition, nor does it account for changes in turbulence regimes acting on a plume due to changes in surface characteristics, for example from land to water or vice versa. The model is not a complex terrain model, although it does include simple algorithms to describe dispersion over slightly hilly terrain. Building effects are managed by simple methods in the model, whereas in reality, aerodynamics in the wake of a building is a complex matter. The primary intent of the building effect algorithm used in OML is to improve concentration estimates applicable for distances beyond about five building heights downwind.

In Sweden, the Swedish Meteorological and Hydrological Institute (SMHI) maintains a model based on the same core code as OML. Like AERMOD and ADMS, OML has been tested against standard US experimental data sets from Prairie Grass, Kincaid and Indianapolis as well as other validation exercises. Results of the evaluations performed with three of these data sets were reported in a European model evaluation exercise in 1994 which used the so-called 'Model Validation Kit' to allow several models to be compared on a similar basis.† Validation will be discussed further in Chapter 9.

HYsplit

The HYsplit (HYbrid Single-Particle Lagrangian Integrated Trajectory) model is a complete system for computing simple air parcel trajectories to complex dispersion and deposition simulations. Version 4 was a result of a joint effort between NOAA and Australia's Bureau of Meteorology to introduce improved advection algorithms, updated stability and dispersion equations, a new graphical user interface and the option to include modules for more chemical transformations.[#]

The dispersion of a pollutant is calculated within a puff or from the dispersal of a fixed number of particles. In the puff model, puffs expand until they exceed the size of the meteorological grid cell (either horizontally or vertically) and then split into several new puffs, each with a share of the pollutant mass. In the particle model, a fixed number of initial particles are advected about the model domain by the wind field. Alternatively, Hysplit4 can assume a puff distribution in the horizontal and use particle dispersion in the vertical direction. In this case, when a particle's size exceeds the grid cell spacing it will split into multiple particle-puffs, each with its respective fraction of the pollutant mass. In this way, the combined advantages of greater accuracy of vertical dispersion parameterization of the particle model and an ever expanding number of particles representing the pollutant distribution as the spatial coverage of the pollutant increases are used.

Other features include the ability to run with multiple nested meteorological grids. Calculations and output fields can be generated on a high-resolution grid and a coarser resolution used when the pollutant expands beyond the domain of the fine grid. Different temporal and spatial resolution of grids may be used. The model can also be run with wet and dry deposition, radioactive decay, re-suspension and with multiple pollutant species.

---

* Olesen, H R and Brown, N (1992, Second edition): *The OML Meteorological Preprocessor – A Software Package for the Preparation of Meteorological Data for Dispersion Models,* MST LUFT-A 122. National Environmental Research Institute, DK-4000 Roskilde, Denmark.

† Olesen, H R, 1995, 'The model validation exercise at Mol: Overview of results. Workshop on Operational Short-range Atmospheric Dispersion Models for Environmental Impact Assessment in Europe', Mol, Belgium, Nov. 1994, *Int. J. Environment and Pollution,* Vol. 5, Nos. 4 – 6, pp. 761–784.

[#] Draxler, R R and Hess G D, 1997, *Description of the Hysplit 4 Modeling System,* NOAA Technical Memorandum ERL ARL-224, December, 24p.

The model has also been tested in a variety of different situations, from inert tracer experiments such as ETEX (Chapter 9), to more complex situations involving real pollutants. Other simulations have included air quality during the Kuwait oil fires.

Access to executable files for HYsplit4 is limited to a version for Windows 95/NT. The model can be run on meteorological data sets available for file transfer. However using the most current meteorological forecast, trajectories may be calculated on-line, as may concentrations.

Improving Visualization of Output

Dispersion models are powerful tools for predicting the transportation, spatial distribution and concentrations of pollutants. However, they can be limited in their data handling, interpolation and visualization abilities. As illustrated in the information boxes earlier in this chapter, the output from a model may not easily be visualized from the tabular output. Many models require considerable amounts of spatially referenced data, (for example large receptor grids, point source emissions, topographic grids) which can be time consuming to prepare.

Geographical information systems (GIS) offer the capability to capture, store, manipulate, analyse and visualize such diverse sets of data. The basic concept of GIS is the location of spatial distribution and relationships. In modelling, the basic concept is one of state, expressed in terms of numbers, concentrations, interaction and dynamics. Integrating the two technologies has many attractions. The two systems may be loosely integrated by running them separately but in a complementary manner through shared data input or data output. Alternatively full integration of a model with GIS, uses a common graphical user interface (GUI) to facilitate the modelling process. The GUI provides a veneer, which assists and guides the user through the modelling process while hiding the processing intricacies. The BREEZE suite offers such integration. Similarly, another commercial version, ISC View carries out all the time-consuming tasks of generating, gridding and displaying plot files for the ISC model. For easy reference and visualization of results, map files in AutoCAD DXF format may be imported and used as a backdrop for contour plots. For report presentation, contour plots may be printed with the overlaid site map or copied to the clipboard and pasted into Windows word processing programs. Such integrated packages eliminate the need for additional graphics packages.

A specific example of this integration is illustrated by a study at Billings, Montana USA.* This area was selected for a number of reasons. The town had the highest annual concentrations of sulphur dioxide in the US in the 1992 National Quality Emissions Trends Report from the EPA. There was much public debate and pressure to clean up the airshed and the city is located in the Yellowstone River valley, framed on three sides by mountain ranges. The climate of this region is affected by a complex topography of foothills, valleys and mountains, with elevation ranges from about 670 m to about 1200 m above sea level. Warm, dry Chinook winds that blow down the side of the mountains move north-eastward through the region. Wind speeds average 4.3 m s^{-1} in August and 5.8 m s^{-1} in December. In the central urban area the average temperature in January is 1.7°C, while temperatures can reach 37°C in the summer months.

Considering the topographic complexities of the area, the level of detail and accuracy desired and the resources available, the ISCST model was chosen. FORTRAN source code for ISCST was obtained from the US EPA's Internet site. Detailed emission information was obtained for each of the six industrial sources in the study area. Sequential surface weather data consisting of hourly stability classes, wind directions, wind speeds, temperatures and atmospheric mixing heights are recorded at the regional airport and were downloaded from the Internet. Land surface elevation and topography data sets (with 30 m grid-cell resolution and a 1:24000 scale) were obtained from the USGS and were combined into a mosaic for the study area. A geographical frame-of-reference

* Folgert J, *Integration of EPA's ISC model with Geographic Information Systems (GIS)*, 2805 SunnyView Lane, Billings, Montana USA jfolgert@globalpositions.com

(receptor-grid) for the area of study was based on TIGER/Line files (Topologically Integrated Geographic Encoding and Referencing).

The complex, irregular nature of pollution concentration demands a range of visualization techniques that may not readily be available. Static maps reflecting the spatial distribution and concentrations may be misleading. One powerful feature of the system is its ability to produce a variety of visualization outputs such as shaded grids, contours (for example Figure 7.3), pseudo three-dimensional viewing and simple animation.

Figure 7.3 Graphical Display of Sulphur Dioxide Concentrations*

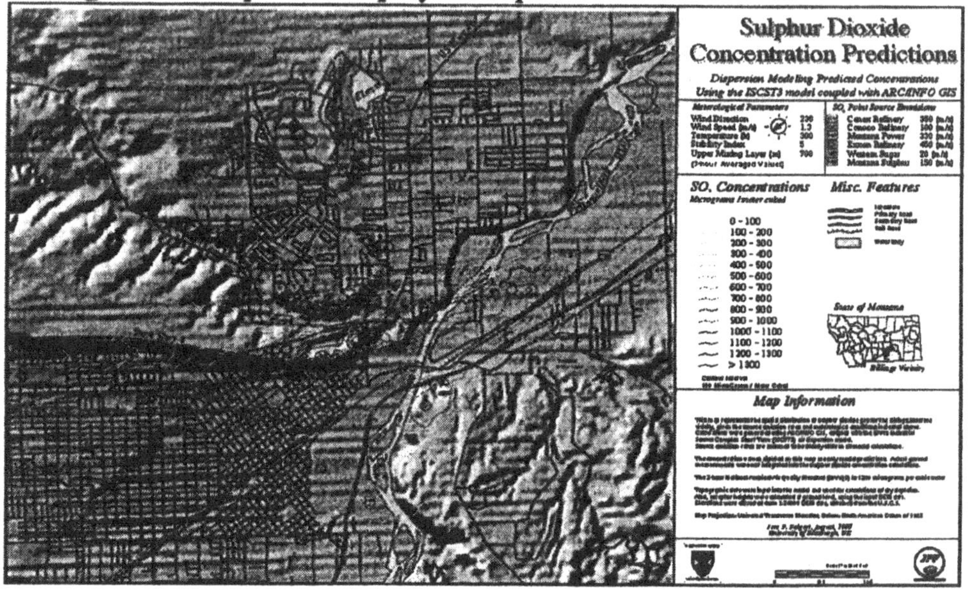

Effective display of air-dispersion modelling results in ways such as this is particularly useful for use in public meetings and for regulatory/policy formulation. Another example is PARAMICS®[†] that can simulate traffic flows along busy networks and represent the resulting pollution levels as a 'web' superimposed over the road system. Such systems are increasingly used for traffic management. Further examples of model output will be given in Chapter 9.

Conclusion

There are conceptual similarities between ADMS and older models such as ISC, as well as simpler models such as R91 and Turner's workbook method. All assume that the plume extends downwind for an indefinite distance, with a 'new' plume being modelled every hour. Refined models may use different approaches according to the scenario. However, computer models for continuous releases and emergency scenarios are more sophisticated in their structure and capabilities than simple models introduced in Chapter 6. They meet specified minimum technical criteria, have undergone field-testing and have been subject to extensive peer review. Older models generally calculate plume spread based on ground level conditions. The latest computer models incorporate increased knowledge of turbulence and diffusion in the planetary boundary layer. Calculations on plume spread are based on conditions occurring at the height of the plume. The models also differ in the calculation of dispersion for sunny (unstable) weather conditions. The vertical spread of pollutant is assumed non-Gaussian with the newer models, compared with Gaussian for the older models. Table 7.2 compares features of the latest models and ISC and some related issues will be mentioned in Chapter 9.

---

* Acknowledgement to J Folgert, 2805 Sunny View Lane, Billings, Montana, USA, jfolgert@globalpositions,com

† http://www.epcc.ed.ac.uk/epcc-projects/PARAMICS/

Table 7.2 Comparison of the Latest Generation of Models with a Predecessor

Feature	ISCST3	Latest generation of model (AERMOD, ADMS)
Source application	All model continuous point, flare, area, line, and volume sources.	
Plume rise	Uses Briggs equations with stack-top wind speed and vertical temperature gradient.	For stable conditions, AERMOD uses Briggs equations with winds and temperature gradient at stack top and half-way to final plume rise. In convective conditions, plume rise is superposed on the displacements caused by random convective velocities. ADMS solves conservation equations to model plume rise rather than using empirical equations such as Briggs.
Meteorological data input	One level of data accepted.	Many data levels are accepted. For example, ADMS allows the use of basic data such as cloud cover, time of day and day of year; or boundary layer data such as the M–O length or surface heat flux and boundary layer height.
Meteorological data in dispersion	Stack-top variables for all downwind distances.	Variables mesured throughout the plume depth (averaged from plume centre line to 2.15 σ_z below centre line; changes with downwind distance).
Plume dispersion: general treatment	Gaussian treatment in both horizontal and vertical.	Gaussian treatment in horizontal and in vertical for stable conditions; non-Gaussian probability density function in vertical for unstable conditions.
Urban treatment	Urban option either on or off. All sources must be modelled using either rural or urban scenarios; no other specification available.	City size and population are specified, so treatment can consider a variety of urban conditions; sources can individually be modelled as rural or urban. The surface characteristics of the modelling domain uses selection by direction and month of roughness length, albedo and Bowen ratio, to provide more flexibility.
Boundary layer parameters	Wind speed, mixing height, and six discrete stability classes only; dispersion curves (P–G) based on surface release experiments.	The description of the atmospheric boundary layer taking account of friction velocity, the sensible heat flux, boundary layer depth, h and the M–O length.
Terrain depiction	Elevation at each receptor point.	The effect of terrain on air flow and hence dispersion is determined using pre-processors that use digitised terrain data.
Plume dispersion: plume growth rates		Uses profiles of vertical and horizontal turbulence (from measurements and/or Planetary Boundary Layer theory); variable with height; uses continuous growth functions rather than a discrete (stability-based) method.
Plume interaction with mixing limit: convective conditions	If plume centre line is above mixing limit or 'lid', a zero ground-level concentration is assumed.	Considers three plume components: a 'direct' plume that is advected to the ground in a downdraught, an 'indirect' plume caught in an updraught that reaches the lid and eventually is brought to the ground, and a plume that penetrates the mixing lid and disperses more slowly in the stable layer aloft (and which can re-enter the mixed layer and disperse to the ground).
Plume interaction with lid: stable conditions	The mixing lid is ignored (it is assumed to be infinitely high).	Considers a mechanically mixed layer near the ground. Plume reflection from an elevated lid is used to account for the effects of sharply reduced turbulence aloft.

Computer models are highly specialized tools. Graphical user interface and data presentation integration make models more accessible and improve visualization of the output, but expertise remains an essential prerequisite to the successful application of models. This need is critical when the more sophisticated models are used or the area being investigated has complicated meteorological or topographical features. A model applied improperly, or with inappropriately chosen data, can lead to serious misjudgements on the source impact or the effectiveness of an air quality management strategy. The resource demands generated by the use of air quality models vary widely depending on the specific application. The resources required depend on the nature of the model and its complexity, the detail of the database, the difficulty of the application and the amount and level of expertise required. Costs may also influence the selection of a model for a specific analysis. These issues are considered in Chapter 9, together with assessments of how models match reality.

Exercise

Taking account of the AERMOD prediction on page 79 for the second highest sulphur dioxide concentration, what is likely to be the hourly average concentration?

CHAPTER 8
MODELLING EMERGENCY RELEASES

Under major hazard regulations, those responsible for installations with the potential for generating major hazards must assess those hazards and formally report to the competent authority. The type of problem at issue is likely to involve the loss of containment and the resulting release of a hazardous substance to atmosphere. The nature of the hazards can be demonstrated perhaps explicitly by a couple of examples. Flammable gas clouds can cause devastating explosions if ignition sources are present. For example, the 30 tonnes of cyclohexane estimated to have been released in the Flixborough incident has been calculated as equivalent to that from the explosion of 16 tonnes of TNT. Even in the absence of an explosion, the thermal radiation from a burning cloud can cause severe damage at some distance. The fireball from 50 tonnes of LPG can injure humans at 500 m.

In this section the concern is with predicting concentrations resulting from accidental releases in terms of relationship to toxic concentrations or those likely to result in an explosion or fire. Often the need is to estimate the distance at which dilution to a safe level occurs. This limit may be defined as the lower flammable limit or a toxicity level.

Just as with conventional air pollution dispersion, modelling accidental releases involves many uncertainties and different types of problem. First a few words about toxicity effects. Exposure to toxic substances involves consideration of both concentration and time. For acutely toxic substances (gases in the present case) it generally is the case that concentration is more significant than time, ie a high concentration for a short time is likely to have a more severe effect than a low concentration for a longer time. Dispersion models predict time-averaged concentrations, whereas toxicity relationships are derived from exposure to steady concentration. Variations in environmental concentrations over time scales shorter than the exposure period in toxicity studies cannot simply be averaged since the effect of peak concentration may be greater than steady concentrations with the same mean value. This is a similar issue to that raised previously on averaging time. While it is important to be aware of these points, correcting for them is not justifiable given the overall uncertainties in both toxicity assessment and in modelling.

Modelling continuous sources involves sources that release pollutants continually or over a period longer than it takes for pollution to reach a receptor. Emergencies often result from very short term releases, even instantaneous releases, such as from an explosion or rupture of a vessel. Concentrations at points downwind from the source depend very much on the time interval after the release and the diffusion along three dimensions. The trajectory is also important but can be difficult to predict. However, this may be less of a problem if radial distances of potential impact rather than exact location are needed.

The idea of a puff release was introduced in Chapter 5 where the solution for the dispersion equation was developed. Taking account of reflection as in the case of continuous sources, a version of the Gaussian puff model used by Turner gives the equation for estimating concentrations downwind as:

$$C = \frac{2Q}{(2\pi)^{1.5}\,\sigma_x\sigma_y\sigma_z}\exp\left[-\frac{1}{2}\left(\frac{x-ut}{\sigma_x}\right)^2\right]\exp\left[-\frac{1}{2}\left(\frac{H}{\sigma_z}\right)^2\right]\exp\left[-\frac{1}{2}\left(\frac{y}{\sigma_y}\right)^2\right]$$

Short term releases, or 'puffs' may only last for seconds, but are of great practical importance from a public health and safety point of view. Unfortunately, it can be very difficult to determine the path or trajectory of the puff, although often the main concern is the magnitude of concentration or the travel time of the puff, rather than the exact point at which the concentration occurs. There is assumed to be no dilution in the downward direction brought about by the wind, which merely translocates the puff.

The puff equation differs from the Gaussian plume equation most notably by having an additional horizontal diffusion term with the consequent disappearance of the wind speed in the denominator of the equation. In general σ_x may be similar to σ_y. The values of σ are not necessarily those used for continuous sources and refer to dispersion statistics following motion of the expanding puff. Table 5.2 indicated typical values. Just as with continuous releases, simplifications are possible for concentrations at ground level and under the dispersion path along the x-axis.

It is often of interest to know the cloud boundary defined by a fixed concentration, which may be the flammability limits, an air quality standard or some other criterion. The line joining points of equal concentration around a cloud boundary is called an isopleth. The procedure for this involves defining the specified concentration of concern $\langle C^* \rangle$ and the wind speed and time after release. The concentrations $\langle C \rangle(x,0,0,t)$ along the x-axis may be calculated by defining the boundary of the cloud along the x-axis. The isopleths at ground level are determined by dividing the equation for centre-line concentration by the equation for general ground level concentration. This equation is solved for y:

$$y = \sigma_y \sqrt{2 \ln\left(\frac{\langle C \rangle(x,0,0,t)}{\langle C \rangle(x,y,0,t)}\right)}$$

The value of $\langle C \rangle(x,y,0,t)$ is set equal to the specified concentration of concern $\langle C^* \rangle$ to give the values of y at each centre line point. This procedure may be repeated for each value of t required. The y-coordinates of isopleths from the x-axis for a dispersing plume may be found using a similar procedure.*

Problems with the application of dispersion models to emergencies are that detailed data may not be available for accident events, and the long computing-run times for complex models may be excessive for providing information in emergency response situations. Nevertheless, modelling can determine the vulnerability zone around a possible hazard. This may be a circle around a stationary site or a corridor along a transportation route. After estimating the zone, other critical factors can be identified, such as:

- the number of people potentially exposed to harm, whether at work, at home or at leisure;
- locations where sensitive people may be found, such as hospitals, homes or schools; and
- access by emergency services, including safe routes and approach distance.

Just as computer models have been developed for continuous sources, so there are many for emergency releases. Some are described later in the chapter and in the appendices. Such models incorporate algorithms for dealing with peculiarities of dense gases and other emergency scenarios. The behaviour of all types of gases can be unpredictable when released in the vicinity of complex terrain, such as narrow valleys or in urban areas. The dispersion of dense gases is also strongly affected by wind speed and atmospheric turbulence. It was described earlier how the path that a buoyant or neutrally buoyant gas takes is mainly influenced by the prevailing wind. Under steady wind conditions the material would move in a straight line directly down-wind, slowly spreading out as it moves. However, wind is not often steady in terms of either direction or speed. Consequently, a gas plume 'meanders' as it drifts. The initial movement of a dense gas cloud is generally influenced more by gravity than by the wind. However, once the momentum of this initial gravity effect has reduced, the prevailing wind begins to influence the movement of the gas. Such features are taken account of in models that have been developed for dense gases and emergency releases.

* Turner D B, 1970, *Workbook of Atmospheric Dispersion Estimates*, First Edition, US Public Health Service Publication No. 26, p 10, (Second Edition 1994, Lewis).

Computer Modelling for Emergency Releases

There are several computer models available for dealing with emergency releases. Examples include ADAM, DEGADIS, SLAB AFTOX etc and summary details of several of these models are given later and in Appendix 3. These models may be used to develop emergency response plans, to train personnel in emergency response procedures and to assess real-world situations rapidly. They also provide valuable tools to support the efforts of organizations manufacturing, storing or transporting hazardous substances to prevent harm from accidental releases. Planners can also use such models to evaluate existing sites and transportation routes, and to design the optimum layout for a new site to minimize risks to its surroundings. Accidental chemical releases may occur anywhere chemicals are stored or used, as well as during transport by road, rail, air or water.

Models for dealing with emergency releases are available in the public domain and commercially. They vary in their features and capability, but share some similarities. The input to these models are often based on a data base of chemicals included in the software, and some include a GUI for ease of use. Like models for continuous releases, meteorological data are needed, and several source types are considered. Figure 8.1 illustrates typical input and outcomes from these models. In such a scheme, the dispersion coefficients may be determined from simple stability input or from more advanced meteorological data.

Figure 8.1 Typical Flow of Data in an Emergency Release Model

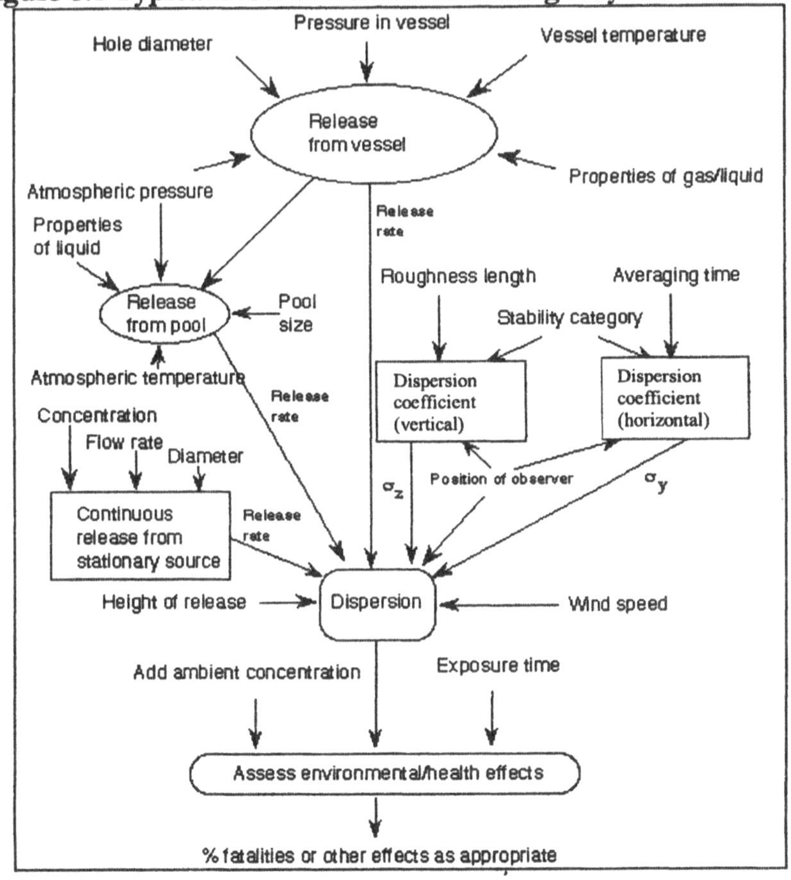

An example of commercial software is GASTAR, developed to model the consequences of dense and neutrally buoyant releases by CERC, the organization that developed ADMS. It is a tool for both land use and emergency response planning. This model uses Monin–Obukhov length or Pasquill–Gifford stability categories. Other essential information includes wind speed and bearing, air and surface temperatures, atmospheric pressure, relative humidity and roughness length.

Another commercial model is PHAST (Process Hazard Analysis Software Tools) developed by Det Norske Veritas Ltd to identify situations that have the potential to cause unacceptable consequences to the population and the environment. This model is Windows based and examines the progress of a potential accident from initial release through formation of a cloud or pool to its dispersion. Figure 8.2 illustrates a screen from the model.

Figure 8.2 Process Hazard Analysis Software Tool (PHAST)

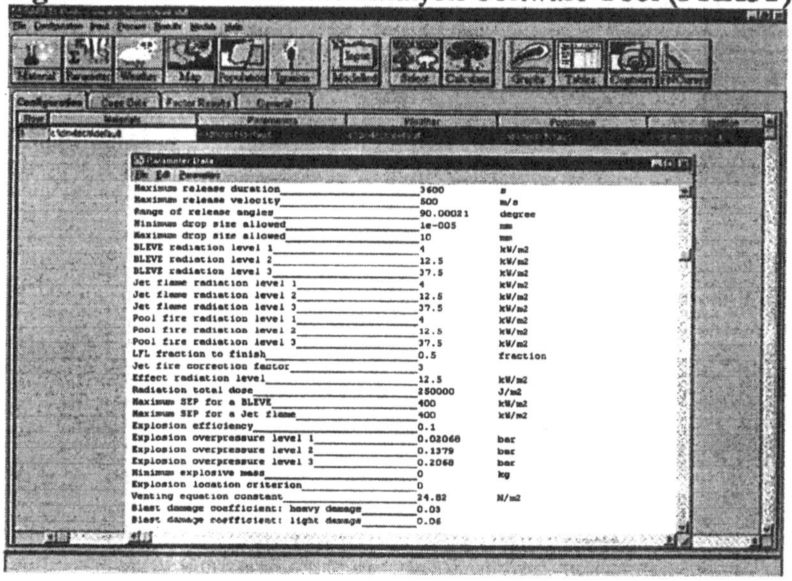

A third example of a commercial model is 'Complex Hazardous Air Release Model' or CHARM, which was originally developed by Radian Corporation of Austin, Texas in the US. It provides a quick and accurate way of assessing the potential impacts of airborne releases and incorporates many of the features described in individual models just outlined. An emergency response team dealing with a chemical spill can use a model such as CHARM in a portable PC and run it on site. The database of over 100 common chemicals contains default data which when combined with estimates of meteorological conditions allows the user to ascertain rapidly the geographical areas at risk, and so concentrate evacuation efforts on those areas. Several scenarios and modes of output are from the model are illustrated later.

When used by the organization storing or using chemicals, models such as these allow the user to develop up-to-date scenarios about potential spills combined with on-line meteorological data. In this way, if a release does occur, rapid response is assured using the best information available. The models use several mathematical algorithms to simulate the instantaneous or continuous release of liquids and gases from contained or unconfined sources. Releases may occur from tanks, from pipelines or from surfaces, such as when a chemical is contained within a bunded area. The release may be a liquid pool fire, a boiling liquid expanding vapour explosion (BLEVE), or a jet fire. Overpressures for a flammable vapour cloud or mechanical failure of a pressurized vessel may also be predicted.

Taking CHARM as an example, this is a puff model that considers any release to be a series of puffs, each of which can be described somewhat independently. The model treats temporal variations in emission rate by approximating a continuous release as a series of small, discrete puffs. Each puff is considered as a symmetrical cloud of contaminant, with advection due to the prevailing wind and dispersion by atmospheric mixing. The model uses a Gaussian distribution to describe the concentration within a puff. The user starts by defining the release and meteorological conditions at the release site. The model takes this input and determines the characteristics of the release and the number of puffs based on the thermodynamic properties of the release species. A typical input file is represented in the following box. This considers a scenario in which there is a release of chlorine from a water treatment works. Notice that the puff model includes a time parameter. Figure 8.3 illustrates an output screen for this scenario representing the situation five minutes after the release.

Version 8.0			
Title:	Release of Chlorine from Water Treatment Unit		
Species:	Chlorine		
Release type: Container/Surface Description			
Emergency response output:		Plume	
X location	-200 m	Y location:	-100 m
Isopleth concentrations (μg/m³):		1.5 , 15 , 1500	
Building height:	60 m	Width	60 m
Source to building distance:	3 m	Direction:	270°
Horizontal tank or pipeline			
Tank length:	3 m	Tank Diameter:	3 m
Liquid depth in container:		3 m	
Tank will not vent during pool fire			
Storage temperature assumed to be ambient			
Storage pressure (calculated, gauge):		6.907 atms	
Droplet fraction of mass same as flash fraction			
BLEVE burn efficiency (calculated):		0.2515	
Hole height above ground	1 m	Hole diameter:	0.1 cm
Hole is facing:	30°	Vertical angle of hole:	0°
Height of hole from container bottom:		0.5 m	
Liquid pool surface height above ground:		0 m	
Maximum pool diameter:	8 m		
Connecting pipe			
Water fraction of spill surface (0 -> 1):		0.15	
Surface is assumed to be concrete			
Surface specific heat:	900 joules/kg K	Thermal conductivity:	1.2 w/m K
Surface density:	2100 kg/m³	Relative pore volume of surface (0 ->1):	0.16
Darcy constant of surface:	0 m		
Meteorological data			
Relative humidity	50 %	Ambient temperature:	25 °C
Ambient pressure:	1013 mb	Cloud cover (tenths)	0
Stability class:	C (Calculated, Sunshine = high)	Solar radiation	0.6286 kW/m²
No inversion present			
Surface roughness	1 cm	Wind measurement altitude	10 m
Winds	Time	Direction	Speed
	00:00	220°	5 m/s
	00:05	230°	5 m/s
	00:10	240°	5 m/s
	00:20	250°	5 m/s

Figure 8.3 Release of Chlorine from a Water Treatment Works

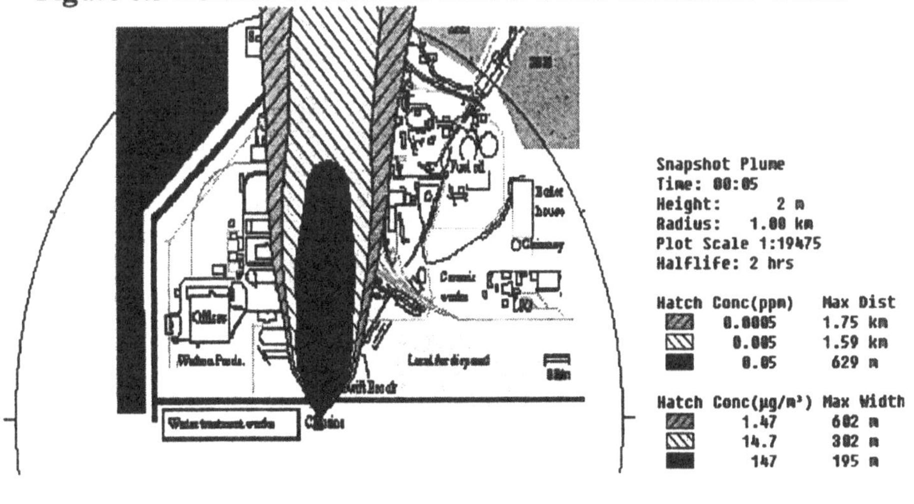

Models such as CHARM can deal with a container/surface liquid release; depressurization; pool/lagoon release and user-specified releases. Algorithms to account for the various processes taking place. Releases from a container can comprise up to four steps: liquid release, depressurization to liquid/vapour equilibrium pressure, boiling and depressurization to atmospheric pressure. If there is no liquid in a container, the first three steps do not occur. If the liquid is stored at a pressure in equilibrium with its storage temperature, the second step does not occur. As the liquid is released, some may form a pool that subsequently evaporates.

In a pool/lagoon release, the model assumes that the pool has a constant diameter and no mass is added to it as evaporation occurs. Effectively this means that the pool-type release assumes that there is no container or mass source for the pool. The pool description should be used only for low volatility liquids. Three processes are allowed for adding energy to the pool to allow it to evaporate. These are heat transfer, mass transfer and solar radiation.

A user-specified release allows the user to supply most parameters for the transport/dispersion model. The inputs required include puff temperature, diameter and depth, puff direction of movement, horizontal puff speed, vertical puff speed, droplet mass fraction, molar water vapour fraction and molar air fraction. A continuous or instantaneous release must be specified.

To fully appreciate the different behaviour of dense gases it is necessary to have a basic understanding of the behaviour of gases. The two major factors that affect the way in which a gas disperses in the atmosphere are the density of the gas and the meteorological conditions during its release. Gases may be categorized by density as being either buoyant (light), neutrally buoyant or dense (heavy). Buoyant gases are lighter than air and so generally disperse upwards away from the source of the release in the manner considered so far. Buoyant gases are very rarely a cause for concern unless the release occurs in stable weather conditions. Neutrally buoyant gases have a density that is similar to that of air. Consequently, the dispersion of these gases is greatly influenced by the meteorological conditions during and after release. In most weather conditions, a neutrally buoyant toxic or flammable gas release can cause hazardous conditions over large distances at ground level.

Dense gases can pose a significant hazard in many scenarios. Initially, due to the effects of gravity, a dense gas is likely to spread equal distances radially around the site of the release to form a 'pancake' cloud at ground level. This cloud may remain at ground level, and the gas at high concentrations, for long periods of time. Dense gases are also likely to find their way into basements and underground culverts where problems of confined spaces occur. Accidental releases of chemicals during transport are most likely to involve dense gases.

Modelling releases that are denser than air, neutrally buoyant or positively buoyant. depends on assumptions made in deriving formulae in the model based on thermodynamic, mass, momentum and energy conservation laws. They include ideal gas behaviour for both the puff and the ambient atmosphere; uniform temperature, density etc within the puff; well-mixed gas and air (ie the puff or plume does not separate) and no heat exchange between the puff and the surrounding environment except by turbulent entrainment. Instantaneous mixing throughout the puff and no variation of dew point temperature or humidity with altitude are further assumptions.

Model output options available at any time depend on the release scenario and the chemical involved. Figure 8.3 illustrated one output. Other options for data output include time-series plots of concentration and dose at a particular location (Figure 8.4) and a time-series plot of the release. Examples of public domain software include ADAM, ALOHA, DEGADIS, SLAB and AFTOX, although commercial firms also offer versions of these models.

Public Domain Software

Examples of public domain softward include ADAM, ALOHA, DEGADIS, SLAB and AFTOX, although commercial firms also offer versions of these models.

ADAM allows the user to specify the source-term right after a release or select from a menu of source-term options for either an instantaneous release or a steady-state

Figure 8.4 Dose Plot at Receptor Location After Ammonia Release from a Fertilizer Works

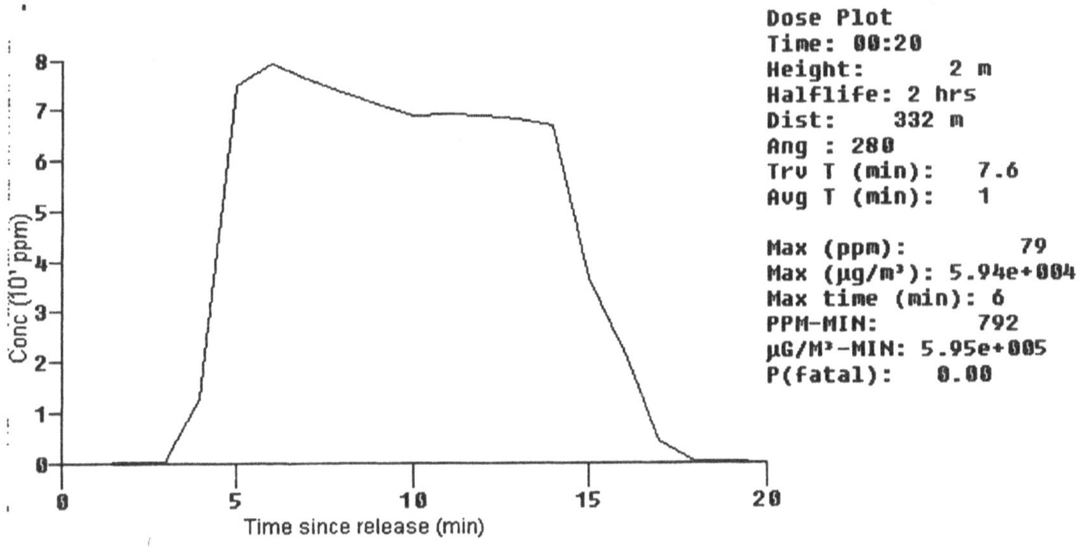

continuous release. However, ADAM can neither handle a limited-duration release nor is it intended for elevated releases. The model contains an algorithm for dealing with releases from jets, which it treats as being on the ground, and horizontally aligned with the wind. ADAM has a database of chemicals from which the user selects the chemical input for the model run. The user also specifies the concentration isopleth of interest, and the averaging time. ADAM will generate a single isopleth showing the location of the pollutant cloud at the specified concentration level, along with a table showing the amount of time travelled by the cloud, its velocity, and the width of the contour. The model will provide peak concentrations as a function of distance and, for instantaneous releases, a peak dose may be calculated as a function of distance for the specified averaging time. In both cases, the information is calculated for the centre line of the isopleth. It is not possible to determine a time-history of concentration at specific points.

ALOHA is a menu-driven model with a graphical user interface. The user selects input for a model run from five menus: chemical, site data, atmospheric, source and computational. The release type may be selected from three options in the source menu: pool evaporation; liquid or gas leaks from a tank; or a gas leak from a ruptured pipe. It is not necessary to enter the release type if the amount of chemical entering the atmosphere as a gas is known, along with other information, such as its temperature. If the chemical desired is in the database, only its name need be specified; otherwise, all the data needed can be entered from the chemical menu. The model generates a text summary, a plot of the footprint of concentrations that are greater than a user-specified concentration, time-series plots of concentration and dose at a particular location, and a time-series plot of the release site. ALOHA includes an embedded version of DEGADIS. Both models have similar formulation but use different numerical techniques to solve the equations and generate different output formats.

DEGADIS (derived from DEnse GAs DISpersion) itself was developed in 1985 for the US Coast Guard. It is an integral 'box' model enhanced with air entrainment and heat transfer algorithms to model heavier-than-air releases at ground level in flat terrain. It can simulate instantaneous, steady state and turbulent releases. Predictions by DEGADIS have been compared against various field tests of heavy gas releases including Canvey Island and Maplin Sands. These included large instantaneous releases of LPG and Freons under meteorological conditions ranging from Pasquill stability classes B to F and wind speeds of 1.8 to 9 m s^{-1}. The model can be applied to a wide range of releases including: gases and aerosols; continuous, instantaneous, finite-duration and time-variant releases; ground-level, low-momentum area releases, and ground-level or elevated upwardly-directed stack jet releases. For aerosol releases, the user has to provide the density of the release, because the model is not capable of doing so itself. The model only simulates one set of meteorological conditions and does not accept any type of real-time meteorological data. No chemical database is provided and so input, including chemical properties, must be entered for each run interactively, or from a file. Terrain is assumed to be flat and unobstructed. Input required includes emission rate; release area and release duration; chemical characteristics; stack parameters; and standard meteorological data. Receptor input includes desired averaging time; height of receptors above ground; and the maximum distance between receptors.

DEGADIS automatically writes an output file. The contents of the file are: input data; plume centre line elevation, mole fraction, concentration, density and temperature at each downwind distance; σ_y and σ_z values at each specified downwind distance; off-centre line distances for two specified concentrations at a user-specified receptor height at each specified downwind distance; and concentration vs time histories for finite-duration releases. A Windows-based version, DEGATEC, is available as well as versions from companies such as Trinity Consultants Inc.

SLAB models denser-than-air releases. It can be applied to a ground level evaporating pool, an elevated horizontal jet, a stack or elevated vertical jet, and an instantaneous volume source. All sources except an evaporating pool may be characterised as aerosols. The model can simulate multiple sets of meteorological conditions in a single run, but it does not accept real time meteorological data. Data are entered directly into the model from an external file. Input data include source type, source properties, spill properties, field properties, and standard meteorological parameters. There is no chemical database, but some chemical properties are available in the user's guide. The model does not generate any type of graphical output, but automatically sends the tabular results to a printer file. These results include: input data; instantaneous spatially-average cloud parameters; time-averaged cloud parameters; and time-averaged concentration values at the plume centerline and at five off-centre line distances at four user-specified heights and at the height of the plume.

Air Force Toxic Chemical Dispersion Model or AFTOX predicts toxic pollutant concentrations from continuous and instantaneous point and area sources. Figure 8.5 illustrates one form of output from AFTOX in the form of a toxic corridor.

Figure 8.5 Example of AFTOX Output for a Continuous Release

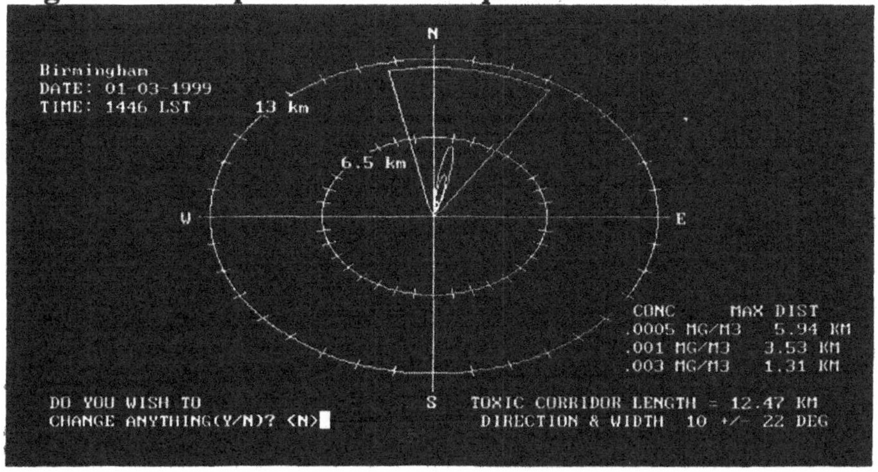

The model was run using similar input data as those for the SCREEN model in Chapter 6.

An alternative form of output from AFTOX predicted a maximum concentration of 3.78 $\mu g. m^{-3}$ at 840 m from the source. This is further away than predicted by SCREEN, but is not unexpected since AFTOX took no account of buildings, and so ignored the cavity effect described in Chapter 3. The AFTOX model calculated for an averaging time of 15 min as the user-selected option. These differences in model output confirm the need for users to be aware of the significance of model input and output if meaningful interpretations are to be made.

AFTOX can also calculate evaporation from a liquid spill and the subsequent transport and dispersion of the vapour cloud.

Conclusion

Many premises produce, store or use bulk quantities of toxic or flammable materials, which in turn may be transported to or from the site. Consequently, there is always a risk of accidental release of these materials. Given appropriate weather conditions, the release may spread over a wide area. Operators of plant, planners, regulators and emergency services all need to be able to predict the consequences of unplanned releases with the potential to cause harm. Critical questions linked to modelling these releases are:

What is the nature of the release (density of gas etc)?
Where will the release go?
When will the release reach critical receptors?
How severe are the impacts likely to be?

Answers to these questions may be given by modelling using the Gaussian puff model. Many public domain and commercial versions of software for emergency release modelling are available.

Exercises

1. A pipeline at a chemical works develops a leak and calculations suggest that the release rate of material will be 375 g s^{-1}. Assuming that emergency procedures operate effectively to shut off the release after 2 minutes, determine the impact five minutes after the release at a location 1 km downwind and 0.1 km perpendicular to the centre line under the resulting plume. The release occurs at night time at a height of 40 m, the wind velocity is 4 m s^{-1} and conditions are overcast.
2. When carrying out worst-case assessments for a toxic gas release, US regulations require that the scenario reflect that the entire quantity stored be released over 10 min. The US emergency response planning guidelines specify 'the maximum airborne concentration below which it is believed that nearly all individuals could be exposed for up to 1 hr without experiencing or developing irreversible or other serious health effects ...'. Consider the implications of these requirements.
3. A factory is situated on the edge of a village about 1 km south and 0.1 km west of a residential area. The wind is blowing at a velocity 4 m.s^{-1} from the south on an overcast day during which a storage tank of chlorine gas ruptures releasing an estimated 400 g.s^{-1} over a period of two minutes. The tank is 30 m high. How much time is available to warn the residents of the problem and what is the likely concentration at the residential area? Assess whether this concentration is likely to present a risk to health.
4. Identify four characteristics of emergency releases that differentiate them from continuous releases in the context of modelling.
5. On a thinly overcast night with a 10 m wind speed of 4 m s^{-1}, a process line 400 m inside the boundary of a chemical works suddenly ruptures. This allows the continuous release of a volatile hydrocarbon vapour at an estimated rate of 25 g s^{-1}. The leak takes place at an effective height of 50 m above ground and 850 m upwind of a nearby farm on the edge of a village some 150 m beyond. A school is located at the edge of the village and 200 m at right angles to the axis of the wind direction blowing from the source to the village. The following day the winds are lighter (3.5 m s^{-1}) and the weather is sunny. With the intention of making a by-pass to the feed to the ruptured pipeline, an operator inadvertently increases the feed to the plant. This in turn leads to an increase in the release to 125 g s^{-1}, and the effective height is increased to 75 m.

   (i) Estimate the steady-state ground level concentration at the closest edge of the village during the night-time release.

(ii) Estimate the concentration at the same time and at a similar distance downwind, but at the location of the school.
(iii) Tabulate or plot the ground level concentration predicted the following day as a function of the distance from the site boundary to a distance of 5 km downwind and directly under the plume.
(iv) Identify the location of the maximum ground level concentration and predict its magnitude.
(v) Assuming that no ambient air quality standards exist for the released substance, what criteria may help in assessing the significance of the emission?

6. Consider a scenario in which a release of flammable liquid from a tank to a pool is followed by ignition of the vapour at ground level. The outcome is modelled using the following input data:

Species: 3-Chloropropene (familiar name allyl chloride)

Release type	Pool/Lagoon	Surface specific heat		800 J kg^{-1} K^{-1}
Emergency response output:	Plume	Surface thermal conductivity		0.32 W m^{-1} K^{-1}
X location	280 m	Surface density		1640 kg m^{-3}
Y location	40 m	Relative pore volume of surface (0 →1)		0.34
Isopleth concentrations (ppm):	0.5 , 5 , 50	Darcy constant of surface		2.8e-007 m
Building height:	0 m	Meteorological data		
Building width	0 m	Relative humidity		50%
Source to building distance	0 m	Ambient temperature		27°C
Direction	0°	Ambient pressure		1013 mb
Liquid pool surface height above ground	0 m	Cloud cover (tenths)		10
Amount released	100 litres	Stability class		D
Pool area	16 m^2	No inversion present		
Pool depth (calculated)	0.625 cm	Surface roughness		1 cm
Wetted tank surface area (liquid pool fire)	0 m^2	Wind measurement height		10 m
Water fraction of spill surface (0→1)	0.15	Winds		
		Time	Direction	Speed m s^{-1}
		00:00	180°	2
		00:05	190°	2
		00:10	190°	2
		00:20	200°	2

Surface is assumed to be dirt.

Pool temperature is assumed to be ambient or boiling point.

The model output is represented in Figure 8.6 over part of the site map. The source is at the centre of the circle and the bold lines represent buildings on the site. Interpret the prediction shown in the figure.

Figure 8.6 Model Scenario Following Ignition of a Pool of Flammable Vapour at Ground Level

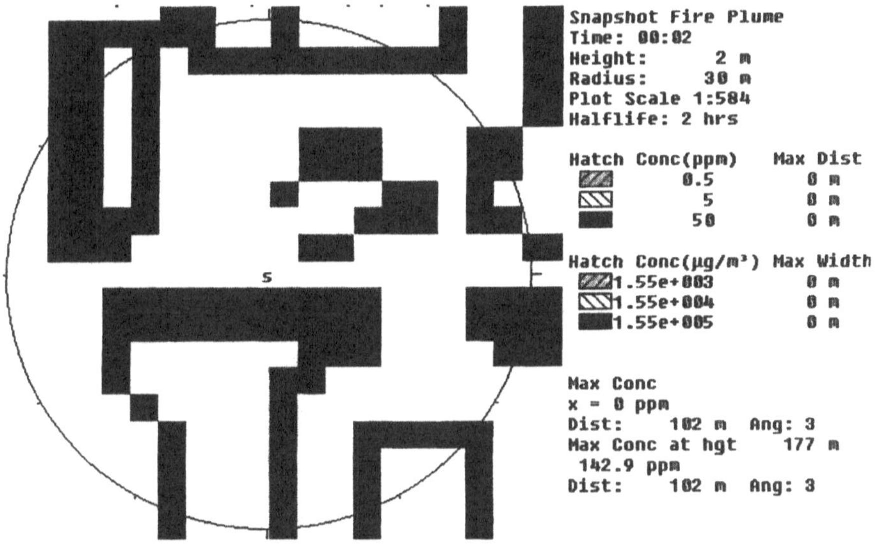

CHAPTER 9
SOME PRACTICAL APPLICATIONS

This chapter draws together some of the issues raised in previous chapters and illustrates their practical relevance to real modelling. However, a starting point is to ask the question of how models match reality in the context of the validation of models.

Intrinsic Uncertainty of Concentration in Modelling

It is important to recognize a limitation in atmospheric dispersion modelling implicit in the discussion on pages 48 and 49 in Chapter 5. The behaviour of components in the atmosphere is governed by the behaviour of the atmosphere itself in terms of meteorological factors.

The application of dispersion models requires meteorological input, and this may be from fairly general assumptions or from detailed meteorological models based on detailed measurement such as those used by the Meteorological Office. Such data are devised from measurements all over the world, which are shared and input to meteorological models. However, there is an inevitable uncertainty in, for example, the estimation of wind speed. Meteorological models have a large unresolved portion u', which is often of the same order of magnitude as u. Therefore, the term <c'u'> is very large and contains most of the turbulent atmospheric diffusion eddies, whose dispersion effects are orders of magnitude larger than the molecular ones. Even with a perfect meteorological model providing detailed information about u, the spatial and temporal scales of the smaller turbulent eddies are so small that a correct numerical integration of the fundamental equation in Chapter 5 would be practically impossible.

The understanding of u' as an unresolvable component that can be minimized but never eliminated is the key to understanding the significance of ensemble averaging. Since u' is a stochastic variable there exists an infinite family of functions u' that satisfies the equation of motion. Thus each possible member u' of the family generates a different concentration c. The average, at a certain point and time, of all possible concentrations generated by the different u' gives the 'theoretical' ensemble mean <c>. The ability to measure u' and c continuously and everywhere, would enable evaluation of the exact member of the family that has occurred in reality. Without this information, it must be assumed that all theoretically acceptable values of u' are equally possible. Thus, in the best possible conditions, the computation of <c> is possible rather than the actual c.

An important conclusion of above is that the output <c> provided by all Eulerian models is conceptually different from the air quality data gathered from monitoring activities. In fact, monitoring data provide estimates of the actual concentration c (with a certain degree of error associated with the monitoring technique), while model outputs are estimates of <c> (with a certain degree of error because of the input data and the numerical and/or analytical approximations). Therefore, even during ideal conditions of no monitoring and modelling errors, model outputs will still differ from concentration measurements. This is often called the intrinsic uncertainty in dispersion modelling. The EPA Guidelines on Air Quality Modelling suggest that this source of uncertainty alone may be responsible for a typical variation in concentrations of as much as +50%.

Some uncertainty associated with a model and its input conditions may be reduced, but neither models nor databases are perfect. Reducible uncertainties are caused by:

- uncertainties in the input values of the known conditions – emission characteristics and meteorological data;
- errors in measured concentrations that are used to compute the concentration residuals; and
- inadequate model physics and formulation.

The reducible uncertainties can be reduced by more accurate and more representative measurements and better model physics. This is what new generation models aim to address.

A number of studies have been conducted to examine model accuracy, particularly with respect to the reliability of concentrations required for comparison against ambient standards. They confirm that:

- models are more reliable for estimating longer time-averaged concentrations than for estimating short term concentrations at specific locations; and
- models are reasonably reliable in estimating the magnitude of highest concentrations occurring sometime, somewhere within an area, but probably not the exact location.

As an example of the latter, errors in highest estimated concentrations of + 10–40% have been found to be typical, ie certainly well within the often-quoted factor-of-two accuracy that has long been recognized for these models. However, estimates of concentrations that occur at a specific time and site are poorly correlated with actually observed concentrations and are much less reliable. Examples of this will be given later in the chapter. As noted above, poor correlations between paired concentrations at fixed stations may be due to 'reducible' uncertainties in knowledge of the precise plume location and to unquantified inherent uncertainties. For example, Pasquill estimated that, apart from data input errors, maximum ground-level concentrations at a given hour for a point source in flat terrain could be in error by 50% due to these uncertainties. Depending on stability and station location, an uncertainty of 5° to 10° in the measured wind direction that transports the plume can result in concentration errors of 20–70% for a particular time and location. Such uncertainties do not indicate that an estimated concentration does not occur, only that the precise time and locations are in doubt.

Scorer* has pointed out that we now appreciate that complex situations become chaotic after a short time. Chaos arises from the approximations that have to be made. When attempting to predict the dispersion of air pollution, it is necessary to have a forecast of wind structure and other weather parameters, and then compute pollutant concentrations after making assumptions about the mechanisms of dispersion. He asserts that this makes predicting concentrations orders of magnitude more unreliable than predictions of air motions. Almost 20 years before, Beychock† commented that many users confused mathematical precision with accuracy. He questioned aspects of modelling including the derivation of the dispersion coefficients, estimation of plume rise, the meaning of concentration in terms of averaging time as well as many other fundamental assumptions of the Gaussian model. To quantify some uncertainties, he took a scenario in rural terrain with a wind speed at 10 m of 2 m s^{-1} and atmospheric stability as Pasquill type B. Other details were as follows.

Stack height	61 m	Stack gas flow	130 kg s^{-1}
Furnace rating	350 MW	Stack gas temperature	232°C
Sulphur content of fuel	0.6%	Ambient temperature	21°C

The calculated buoyancy factor was 251 m^4s^{-3} and the maximum plume rise was 532 m at 1084 m downwind assuming that the wind speed at the stack top was the same as at 10 m.

Several changes were then explored to reflect uncertainty. Plume rise was increased by 20%, the vertical dispersion coefficient was decreased by 25% and a 10° shift in wind direction was assumed. In addition, the assumption by EPA that Gaussian models yield 1 hr means rather than 10 min means was considered. The outcome of these assumptions is summarised in Table 9.1. The

* Scorer, R S, 1998, 'Modelling of air pollution – its use and limitation', *Clean Air*, 28 (3) pp 102–104

† Beychock, M R, 'How accurate are dispersion predictions?' *Hydrocarbon Processing*, October 1979

over-prediction can be seen to range from 6 to 80 times. Nevertheless, Beychock concluded that so long as it is remembered that models do not necessarily reflect reality, they are useful tools.

Table 9.1 Ground Level Concentrations Calculated Under the Plume Centre Line

	Prediction by Gaussian dispersion model				
	A	B	C	D	
Downwind distance to receptor (km)	Base model*	Increased plume rise and decreased σ_y	B + wind shift correction	C + averaging time correction	Over-prediction (A/D)
2	16.1	1.1	0.5	0.2	80
4	20.5	13.8	6.0	2.4	9
6	12.0	12.0	4.7	1.9	6
8	7.4	8.4	3.1	1.2	6
10	5.0	5.9	2.0	0.8	6

* Base model assumes Briggs plume rise, Gaussian dispersion, power law conversion of surface wind velocity (with exponent 0.15) to determine velocity at stack tip, and concentrations assumed to be 1 hr means.

So what does the concentration really represent? Remembering the significance of averaging time in relation to concentration measured in real time, we may consider the same issue in relation to predicted concentration. Does a modelled value represent an instantaneous concentration or an average over 3 minutes, 15 minutes, 1 hr or 1 year? In general, it is regarded that predictions represent averages over 10–15 minutes, although many US EPA dispersion models yield 1 hr average concentrations, while longer term averages may also be specified. If values for other averaging periods are needed, conversions can be made using appropriate factors as described in Chapter 3.

How Models Match Reality

A typical boiler or furnace on a process plant will be used for comparing several modelling approaches. The release data are presented in Table 9.2.

Table 9.2 Release Data Used for Model Comparison

Release characteristics from furnace	
Release height (m)	80
Diameter (m)	1.5
Efflux velocity (m s^{-1})	15
Exit temperature (°C)	250
Pollutant release rate (g s^{-1})	1

In all cases ground level concentrations were predicted for the simple situation of flat ground. In practice, the influence of buildings and hills should be incorporated. Figure 9.1 shows the predicted maximum ground level concentrations for different weather conditions for the furnace release.

It is clear that ADMS predicts higher concentrations for sunny daytime (unstable) and days or nights with partial cloud (neutral) weather conditions. The difference in the predictions is greatest for the very sunny summer weather conditions. These differences are expected since ADMS uses a non-Gaussian vertical distribution of pollutant for these conditions. This change in methodology has the effect of bringing the plume down to ground quicker and hence closer to the stack. The plume does not spread as much, is diluted less and so the concentration is higher. It will be recalled from Chapter 3 that this prediction matches reality. During night-time, with clear skies (stable conditions), the situation is reversed; ISC predicts higher maximum concentrations. However, for these weather conditions the concentrations are much lower than for the unstable and neutral conditions. Hence, they are not a governing case for determining stack height or pollution abatement requirements.

Figure 9.1 Relative Maximum Ground Level Concentrations Predicted for Different Weather Conditions*

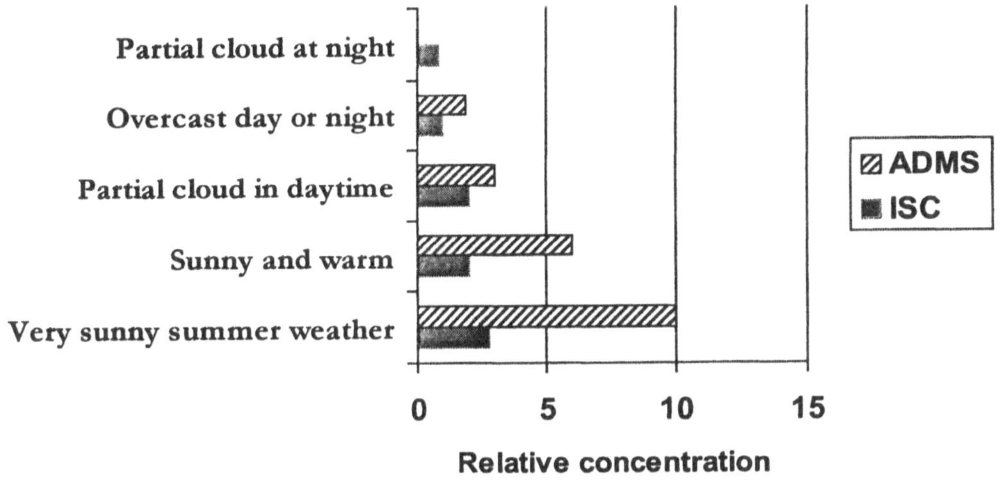

*Adapted from *Environmental Protection Bulletin*, 047, 1997, pp13–17

The findings however are not as clear-cut as the initial review implies. The ADMS model predicts high ground level concentrations closer to the source than ISC but lower concentrations further away. For large industrial sites such as petrochemical plant this would almost certainly suggest that the highest concentrations would appear inside the site boundary and the public exposure would be at lower levels than previously expected. The location of the maximum concentration needs to be further considered. Prediction by putting the simple dispersion equation in mathematical software (MATHCAD) generated the variation as shown in Figure 9.2.

Figure 9.2 Relative Maximum Ground Level Concentrations at Different Distances from the Source Predicted by the Basic Gaussian Plume Equation

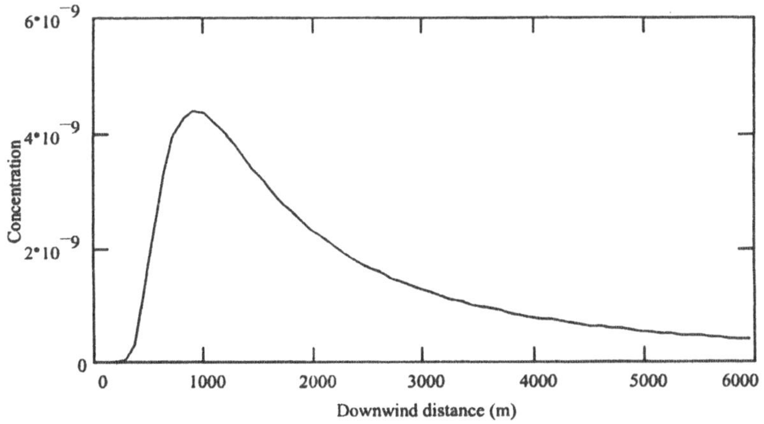

The maximum at 1000 m may be compared with the traditional rule of thumb introduced in Chapter 2 that the maximum occurs at 15 times the stack height from the source. This places it at 1200 m from the source, whereas SCREEN also predicted a maximum at 1000 m.

Figure 9.3 represents the results from more advanced models for similar very sunny summer weather (extremely unstable) conditions. This shows that ADMS predicts that the maximum concentration occurs much closer to the source than the older generation model ISC and simpler models. However, concentrations further away are lower. Table 7.2 gave some reasons for the differences.

Figure 9.3 Relative Maximum Ground Level Concentrations Predicted by ADMS and ISC for Different Distances from the Source*

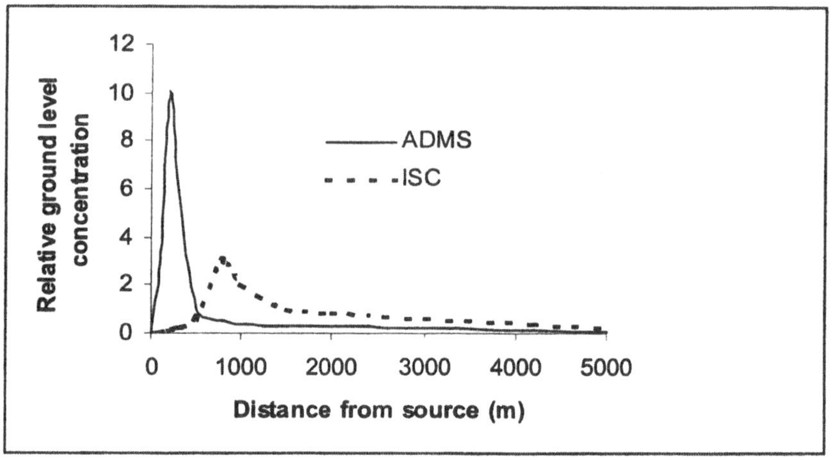

*Adapted from *Environmental Protection Bulletin*, 047, 1997, pp13-17

The predicted concentrations highlight the importance of examining the impact on air quality at locations where sensitive receptors exist. These include nearby residential areas, schools and hospitals. This is particularly so for large process plant where the maximum concentration predicted by ADMS is likely to occur within the site boundary. In this situation, the maximum concentration may be more relevant in respect of occupational exposure rather than general air quality. It will be remembered from Chapter 1 that ambient air quality standards are more demanding than occupational exposure levels to allow for susceptible members of the population.

Other comparisons have found that ADMS predicts higher concentrations than AERMOD and ISCST for releases over 35 m and lower concentrations for releases below 20 m.[*]

Evidence suggests[†] that large errors occur when conventional computer models are used to predict dispersion around buildings. While ADMS may predict higher ground level impact from an elevated point source, AERMOD may underestimate ground level concentrations influenced by building effects. For the flow around buildings. ADMS uses a similar treatment to ISC-PRIME in modelling the building cavity and main wake region. Not surprisingly, therefore, similar results have been found in model comparisons.[‡] ISC3 does not model the cavity and treats all stacks within the zone of influence of a building identically. Results from SCREEN were much more conservative, which is as it should be. Computational fluid dynamics models are based on solutions to equations for flow. Such models allow calculation of building effects and more representative results have been suggested. Under complex topographical conditions, wind tunnel modelling is the most reliable.

This discussion has so far considered how selected modelling systems compare against each other. Validation against monitoring is a separate issue.

---

[*] *Comparison of regulatory design concentrations: ADMS versus AERMOD and ISCST3*, Cambridge Environmental Research Consultants, 3, Kings Parade, Cambridge, September 2000.

[†] Harvey D J, & Obasaju E D, 1999, 'Accuracy of techniques for modelling the effects of building downwash on the dispersion of emissions to atmosphere', *Clean Air*, 29, 42–45

[‡] *Comparison of the modelling of building downwash: ADMS versus ISC3, ISC-PRIME and SCREEN*, Cambridge Environmental Research Consultants, 3, Kings Parade, Cambridge, September 2000.

Validating Models

It should be clear from the foregoing discussion that there are several sources of uncertainties in modelling. There could be data collection errors, especially regarding source-related measurements. An incorrectly sited meteorological instrument may not give representative wind directions and speed. There could be errors in the estimation of the parameters not directly measured, especially for the classification of the stability and plume rise. These sources of errors are in addition to the limitations in the basic theoretical framework of the model described previously in Chapter 5. These sources of uncertainties in the air quality model must be appreciated before embarking on the application of a model.

Performance measures

There is increasing demand for more objective and formalized procedures in order to evaluate the quality (fitness-for-purpose) of models. Evaluation procedures are essentially a consensus of those workers active in a field (including model developers, model users, regulatory bodies etc) as to what is a reasonable and essentially pragmatic strategy for determining model quality and communicating it to interested parties. Interested parties often include less skilled users who themselves are under increased pressure to provide quality assurance.

The development of an evaluation strategy is most obviously linked to the purpose of a model. A proper evaluation strategy follows a consideration of, for instance, the following questions:

- What input variables?
- What output variables?
- How to compare model with measurement?
- What statistical measures to use?
- What is concluded from these statistical measures about the fitness-for-purpose of a model?

Many of these questions are not easily answered. Model evaluation involves scientific assessment, user-oriented assessment, code verification and model validation. An additional factor has to be the resources available and the run time required by models.

Various statistical performance measures are available to compare model predictions with data. These have been used extensively for the series of European Workshops on the Validation of Short Range Dispersion Models.* While no single statistical measure serves to indicate a model's performance, the range of measures provides alternative ways of presenting information and can allow conclusions to be drawn. For a set of observed data and a corresponding set from modelling, the measures include means, standard deviations, bias (B), fractional bias (FB), normalized mean square error (NMSE), correlation (R) and fraction of data within a factor of two (F2).

These measures have been employed in a number of model-evaluation exercises including in European Dispersion Model Harmonization workshops. While the application of the statistical measures is useful for model evaluation, care must be taken in interpreting the results by considering the values of the measures to be expected. For example, the inherent uncertainty was mentioned in previous chapters. Consequently, it is evident that in comparing concentration, values of 0 and 1 for NMSE and R would not be expected. These values would imply a perfect match between data and calculations. After validation, a model becomes more or less site specific. The importance of this in terms of local variations will become clear from the examples later in this chapter.

Validity of a model must be established by judging its performance with respect to a set of observed air quality data. This implies a planned experiment at the project site or in the urban setting where source and meteorological parameters can be collected simultaneously with data from

* Olesen H R, 1997, 'Use of the model validation kit at the Oostende Workshop: overview of results'. *International Journal of Environment & Pollution*, Vol 8 Nos 3/4/5/6.

monitoring sites. While large-scale examples will be described later, a simple example to introduce this aspect is drawn from a study by the author of airborne fluoride carried out around a large brickworks. Measurements of gaseous fluoride in the atmosphere were obtained at ten sites, seven of which were in a line through the source and along the direction of the prevailing wind. The location of the source is shown as the origin in Figure 9.4. The broken line shows the predicted location of the maximum ground level concentration based on an effective source height of 80 m with predominantly D class stability. As shown in Figure 6.3(a), this should occur at about 2.5 km from the source. Figure 9.4 shows that using annual data, the location of the maximum ground level concentration matched that found by monitoring suggesting validity in the approach.

Figure 9.4 Airborne Fluoride Monitoring at Seven Sites Confirms the Maximum Ground Level Concentration at Predicted Location over the Long Term

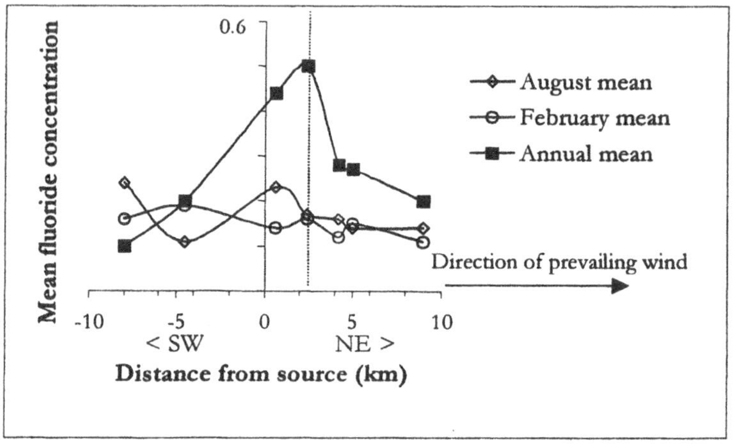

It is also evident from Figure 9.4 that using shorter term mean values there is not a good match against the prediction. It is worth recalling here that air concentration maps produced by models represent an average over a period, rather than a snapshot valid at a particular time. Therefore, when interpreting the output of models, this factor must be borne in mind. Its effect is illustrated by an example that plumes may be larger than the actual plumes as might be seen from an aircraft or in an image from a satellite. The larger size is due to the advection and dispersion processes that occur over the averaging period. The advection or the time the plume spends over a particular point, and the horizontal and vertical mixing will affect concentration levels. Problems with short term values will be developed further later.

More detailed model validation is quite difficult to achieve satisfactorily, but some examples will be given shortly. Such studies can be costly and can still be inconclusive due to the complex nature of atmospheric transportation and dispersion, especially in terms of averaging time and natural variability. Often, therefore, models may be applied without appropriate and documented model validation. On the other hand, some model applications have considered what is called as the uncertainty analysis around the predictions of the model. Uncertainty analysis attempts to establish the likely spread of the model predictions due to errors in the data that include both observed and estimated components.

Modelled pollutant concentration should not be expected to match observed concentrations exactly, even for a matching averaging time. This point was emphasized earlier. The mismatch between the observed and modelled concentrations generally increases when modelling short term pollutant concentrations such as over 10 minutes. For long term averages, however, more encouraging modelling results have been found. So modelling becomes attractive for planning situations rather than for control applications.

Data used for validation should be representative and within the operating range of the model. Several of the standard data sets used for validation have been mentioned in previous chapters. Other validation exercises have taken advantage of appropriate opportunities such as releases from the Chernobyl accident,* or the Kuwait oil fires.

Opportunistic Validation

Following the Chernobyl disaster in 1986, the Meteorological Office began work to develop a model to predict the impact of such incidents. The Nuclear Accident Model, NAME, allows such predictions. NAME is a mesoscale and long range, multiple-particle dispersion model that is the UK national operational model for emergency response in case of a nuclear accident.

Operation Desert Storm in 1991 in Kuwait raised possible consequences of the hypothetical dispersal of chemical warfare agents from chemically armed missiles, the transport and dispersal of radioactive material that might be released by air attacks on nuclear sites and the effects of smoke plumes from burning oil wells. Smoke emissions from the oil wells provided another opportunity to validate the NAME model and to predict the impact of the fires. One concern was that the dense smoke could absorb sunlight, thereby heating the plume and causing it to rise through the atmosphere. If the plumes reached the upper atmosphere, they could be trapped and lead to severe consequences for the global climate. The Meteorological Office collects weather information continuously from human observations, radar and satellite systems worldwide and these data are used to produce global weather forecasts several times daily. The weather data provide meteorological input to the NAME model. While weather satellite images provided information on the position of the plumes, quantitative information on their composition came from C-130 Hercules aircraft flying through the smoke. Peak concentrations of particles exceeded 600 $\mu g\ m^{-3}$ a few hundred metres above ground. The conclusion from modelling with NAME was that the upper atmosphere would not be polluted as feared. Version, 2.1 of NAME was introduced by the Meteorological Office in March 1996 to include improved scavenging coefficients for washout from mixed-phase (ice and water) cloud, and the inclusion of a representation of plume-rise at source.

Another modelling exercise† generated data on a daily basis for numerous US government agencies, the United Nations World Meteorological Organization, and nine Middle Eastern nations. Dispersion of the oil fire smoke plumes was calculated initially for a few specific events during the Gulf War and then daily from the middle of May 1991 until the last oil well fire was extinguished on November 6, 1991. Approximately 4.3 million barrels of oil per day were burning from well heads in early June 1991. Eight oil fields were grouped into three source regions. The fraction of the burning oil that went into particulate smoke (the smoke emission factor) was taken to be 2%.

The principal atmospheric transport and diffusion models used were the Mass-adjusted Three-dimensional Wind (MATHEW) model and the Atmospheric Diffusion Particle-In-Cell (ADPIC) model. The air concentration of smoke in the horizontal plane was calculated at nine levels in the vertical. Deposition of the particles on the ground was also calculated and contours of ground deposition were generated. The atmospheric wind forecast and analysis data were provided by the United States Air Force Global Weather Centre.

The plume from the ADPIC model for May 8, 1991 extended along the Persian Gulf from Kuwait to the south-east beyond Oman over the Arabian Sea. A meteorological satellite image for the same time, showed the actual position of the plume as seen from space to be 100 to 200 km to the west of the model prediction. This position was within the limits of the uncertainty in the input wind data, which had a spatial resolution of 90 km.

The forecast on May 17 showed a significant problem with the modelled wind flow and subsequent plume position. Preliminary investigation of the disparity in the calculated versus real plume location pointed to a lack in detection and analysis of a cold front moving south-eastward through the region. The forecast missed the strong inland penetration of the smoke plume flow over Saudi Arabia. In conjunction with the cold front there was a distinct change in airmass characteristics including a more stable boundary layer and lower and stronger inversion, particularly

* For example, http://www.ei.jrc.it/ei/database/index.html

† For details on ARAC see http://www-ep.es.llnl.gov/www-ep/atm/ARAC/arac.html

during the day. An upper air sounding indicated a capping inversion at about 1000 m. This example illustrates the need for reliable and timely meteorological data that form the basis of modelling.

Urban Validation

In all applications of models an effort is needed to identify the reliability of the model estimates for a particular area and to determine the magnitude and sources of error associated with the use of the model. To improve the basis for decision-making, studies are continuing on procedures for determining the accuracy of models, quantifying the uncertainty and expressing confidence levels in decisions that are made. Several studies have taken place to assess the performance of the latest generation of models in the process of review and assessment of air quality as required under the Environment Act 1995.

One study assessed the performance of the ADMS-Urban model (version 1.50) by comparing model predictions with monitored concentrations of nitrogen oxides (NOx, NO_2) and sulphur dioxide (SO_2) at locations in Central and East London for both a summer and an autumn/winter period. Hourly sequential meteorological data comprised wind speed, wind direction, surface temperature, precipitation rate and cloud cover, together with day of the year and time of day. The data for London Heathrow airport in ADMS format were obtained from the UK Meteorological Office and were used as input to the model. The ADMS meteorological pre-processor calculated the boundary layer height and the Monin–Obukhov length, which were used to derive the hourly boundary layer turbulence profiles required for the dispersion calculation.

While the surface topography varies, a constant value of z_0 = 2m was used to represent the surface roughness of central London. The complex terrain module in ADMS was not used in the study. Other adjustable parameters were set to their default values. An attempt to account for the increase in turbulence due to heating in the urban area was made by adopting a minimum value of Monin–Obukhov length of 100 m in stable conditions.

Aggregated emissions, were supplied on a 1 km square basis, and were modelled as supplied in ADMS-Urban as shallow volume sources of dimensions 1 km (length) x 1 km (breadth) x 100 m (depth). The emissions and stack data consisting of stack height, diameter, gas volume flow rate and exit temperature as well as mass emission rates of pollutants were only available for the processes authorized as Part A in terms of the Environmental Protection Act 1990. The same data were obtained for additional sources in the East Thames corridor. Therefore, only Part A processes were modelled explicitly while it was assumed that most Part B sources were accounted for within the aggregated emissions data. Emissions were assumed constant with all sources operating for 24 hr each day since no data on hourly variations were available for point sources. Road links were modelled explicitly using emission factors from the *Design Manual for Roads and Bridges* mentioned in Chapter 6. The emissions from roads were assumed to vary linearly with the diurnal traffic variation. Seasonal changes in traffic or other emissions were ignored.

ADMS calculated hourly averaged concentrations of SO_2, NOx and NO_2 at each receptor site for every hour in both of the three month study periods (May–July, October–December 1995). This allowed the model data at each site to be compared with the monitored concentrations as a time-paired data series and allowed analysis of the model predictions with weather conditions. Recorded hourly-averaged concentrations of SO_2, NOx and NO_2 from monitoring sites including data from the National Air Quality Archive were used for comparison with the model predictions. The PC used in this study had a 200MHz Pentium-Pro processor, 128Mb of RAM and 500Mb of free hard disk space. Typically it took 2–3 hrs to run the model for each of the three-month periods.

ADMS-Urban can be used in the air quality review process to calculate means and percentiles of pollutant concentrations over a long period (at least one year) allowing assessment of air quality against the objectives in the national strategy. In this study, the investigators therefore decided to compare the percentiles of pollutant concentrations, although the summer and winter study periods were both relatively short (three months each). Comparing predicted mean and 95th percentile of hourly average concentrations of SO_2 with that observed at each monitored site for both the winter and summer generally showed good model performance. Agreement with higher percentiles showed more variability between the seasons and from site to site. For paired time-series of observed and modelled data points, the correlation and fraction of modelled values within a factor of two of the observed values were low. Time series plots suggested reasons for this. Concentrations of SO_2 were generally low with occasional high peaks. While ADMS can predict approximately when peaks occur, the time may be wrong. This was attributed to variability in the weather conditions between the meteorological data site (Heathrow airport) and the easterly extreme of the study area, where the major SO_2 sources are located. There is also the time it takes for pollutant released to the east of London to reach the monitoring sites. This time-related problem has long been recognized as the earlier comments on intrinsic uncertainty suggested. Such uncertainties do not indicate that an estimated concentration does not occur, only that the exact time and locations are in doubt.

Plots of concentration against meteorological variables indicated that ADMS and observations generally showed similar variation with wind speed and Monin–Obukhov length, although ADMS tended to predict higher concentrations for lower boundary layers. There were also differences between ADMS and the observed values for wind direction.

During the summer period high observed values at the monitoring sites in Central and East London occurred when the wind at Heathrow was blowing from the south-east (135°) whereas ADMS predicted higher concentrations for easterly winds (90°). The observed concentrations and the concentrations predicted by ADMS were of the same order of magnitude. This suggested a shift in wind direction across the study area, so that the when the wind was blowing from the south east at Heathrow, it was easterly in the East Thames corridor. At the City and Bloomsbury sites in summer higher maximum concentrations were predicted by ADMS than were monitored. This was attributed to inherent uncertainty (ie concentration fluctuations) or to the emissions being poorly defined through the lack of accurate time-varying emission data. Such source characteristics were highlighted at the beginning of this chapter.

During the winter, ADMS significantly over-predicted concentrations of sulphur dioxide when the wind was blowing from an easterly direction. Analysis of the ADMS predictions showed that these peaks were again caused by the large power stations in the East Thames corridor. As there were no corresponding peaks in observed concentrations for a south-easterly wind direction as occurred during the summer period, this suggested that the large sources were not operating during this period. These results further confirmed the importance of accurate information on time-dependency of emissions from major sources.

As shown in Appendix 1, the National Air Quality Standard for sulphur dioxide has been set at the 99.9th percentile of a 15-minute average. Over a period of a year this corresponds to the ninth highest value calculated by ADMS using hourly sequential meteorological input. Over a period of three months, this is the second highest value calculated. Therefore if ADMS is to be used to perform calculations for this standard, accurate hourly sulphur dioxide emission rates from large sources are needed. Smaller sources can also have a significant local effect on concentrations, and where possible, these should be included explicitly in the emissions database, giving the modeller the option of whether to model them explicitly or as aggregated emissions.

Mean concentrations of NOx predicted by ADMS showed good agreement with monitored mean concentrations, although some over-prediction occurred during the summer. This could be because seasonal variation in emissions was absent from the emissions database and NOx emissions may be greater in winter through increased use of boilers and greater likelihood of catalytic converters on vehicles running inefficiently. Statistics for the time-series pairing of observed and predicted values showed better agreement for NOx than sulphur dioxide for all sites. The NOx results were less sensitive to wind direction than SO_2. This was attributed to sulphur dioxide concentrations being dominated by individual sources located in approximately one direction from all the receptor locations, whereas NOx emissions were more evenly distributed around the receptors. When the wind was blowing from the south-east, ADMS underestimated NOx concentrations at Bexley and Greenwich because NOx emissions to the SE of the sites were ignored. Again, inadequate emission data were at fault.

This study showed ADMS-Urban to be a practical and powerful tool for calculating concentrations of pollutants within a large urban area provided that adequate input data are available. Developments likely to improve model performance include methods for enabling emissions during previous hours to properly impact on current concentrations. Emissions on previous hours are only accounted for in the box-model for calculation of background ozone concentrations. Inclusion of their effect is likely to have a significant influence on calculated concentrations especially when the wind is very light or where emissions are changing rapidly.

The INDIC AirViro system includes several models. The Lagrangian–Gaussian approach is recommended for areas with reasonably flat topography and was used for another study in the London area to assess the capabilities of dispersion models for air quality assessments on a regional scale. The system also includes an Eulerian advection–diffusion grid model, more suited to larger areas with complicated topography, two street canyon models and a heavy gas model.

Comparison of the monitoring and modelling predictions in this study involved long term statistics, ie the 3-month means for both summer and winter, and the 98th or 99th percentiles of the 1 hr means for the 3-month periods. And hour-by-hour modelling and monitoring data for the selected 24 hr periods. Unlike ADMS, but in common with many other atmospheric dispersion models, this model is currently not capable of calculating 15-minute mean concentrations.

Time series measurements of NOx, NO_2 and SO_2 over 24 hr periods were superimposed on modelled pollutant concentrations, and presented as graphs such as Figure 9.5. For NOx data in Central London, the model over-predicted seasonal means by factors of between 3–8, and the 98th percentile by factors of between 4–7. For East Thames corridor stations, the model was accurate for seasonal means to within a factor of 2 and a factor of 3 for the 98th percentile.

The greater degree of over-prediction for Central London was attributed to suspect high NOx emissions specified in the Central London emissions database. As with any modelling system, the accuracy of the predictions depends on the quality of the emissions database. A number of concerns were raised about some emissions data, particularly area source components that did not seem to correspond with NAEI data. The current system was not considered capable of accurately predicting compliance with the NAQS standards without further validation of emissions data.

Figure 9.5 Indic Airviro Dispersion Modelling and Continuous Monitoring Results at Greenwich (10 am May 5 to 10 am May 6)

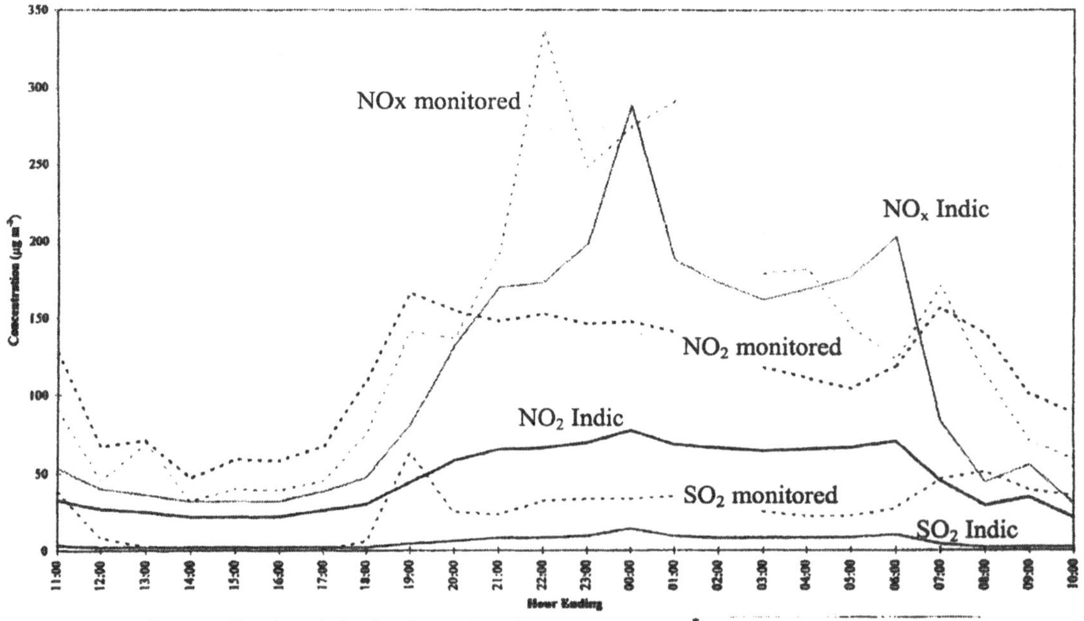

Source: *Benchmark Study, Atmospheric Emissions Modelling**

* *Benchmark Study on Atmospheric Emissions Modelling Air Quality in London.* Report to the London Borough of Croydon, Stanger Science & Environment, December 1997.

It was further concluded that the system in its current format is capable of predicting a range of time-averaged concentrations, ie annual (or seasonal means), and percentile concentrations (at integer intervals) of hourly means. The model was not capable of predicting the 99.9th percentile, although this would be a relatively straightforward modification. In addition, the model could not predict 15 min means, although the uncertainties of any method based upon hourly meteorological data must be considered.

A particular benefit of the model is that it is possible to derive several time-averaged concentrations from a single model run, giving considerable cost savings in time. The UNIX platform on which the model operates may be viewed as a disadvantage, but there are advantages. During routine operation, the number of emissions sources are unlimited, run time is relatively short even for large multi-source scenarios using seasonal data, and several scenario runs may be undertaken simultaneously.

Another study compared predictions from these two models against air quality measured at automatic urban network (AUN) stations in the West Midlands.* AirViro was used to generate maps of predicted concentrations of CO, NOx and PM_{10}. An example of the output shown in Figure 9.6 followed the shape of the built environment and demonstrated the effects of local emissions and arterial roads.

Figure 9.6 Model Output for NOx†

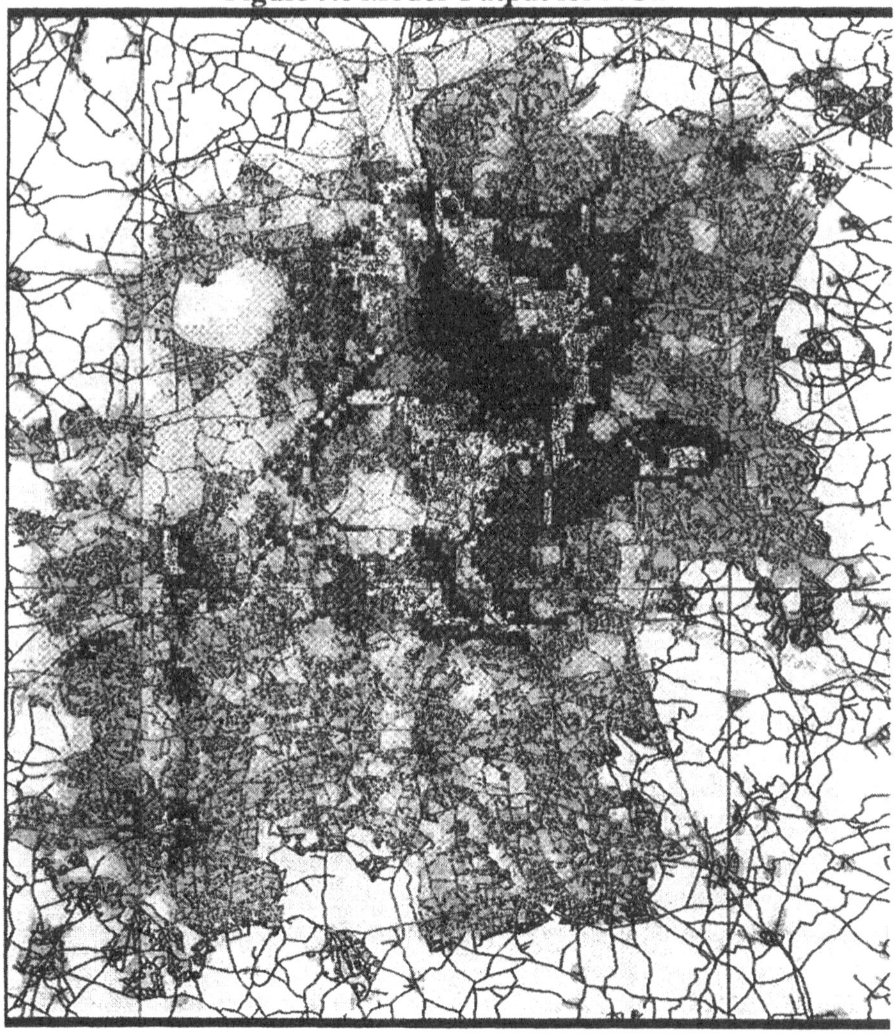

* Appleby R S, *Clean Air*, 28, No 2, pp 64–66.

† With acknowledgement to R S Appleby, Environmental Protection Unit, Birmingham District Council.

When both AirViro and ADMS-urban were run for summer and winter episodes, it was found that both followed the pattern of measured concentrations and predicted many observed peaks. However, both showed inaccuracies, but predicted values within a factor of two or three of the measured values. It was concluded that such models are a cost effective means to determine problem areas in local air quality management.

A Problem with NOx

Nitrogen dioxide (NO_2) and nitric oxide (or nitrogen monoxide, NO) are both oxides of nitrogen that collectively are referred to as NOx . All combustion processes produce some NOx emissions, largely in the form of NO, which is then converted to NO_2, mainly as a result of reaction with ozone in the atmosphere. NOx features prominently in the priorities for local air quality management, yet has special problems owing to the complex chemistry involved in the conversions. Dispersion models can be used to estimate NOx concentrations (NO plus NO_2) directly, but some of the uncertainties have been indicated in the previous section. Corrections are needed to obtain the NO_2 concentration since NOx concentrations can be much greater than NO_2 concentrations in some situations. An empirical relationship between hourly mean NO_2 and NOx concentrations may be applied based on measurements in similar locations. One such model* based on measurements in London gave the relationship:

$$[NO_2] = 2.166 - [NOx]\{1.236 - 3.348(\log_{10}[NOx]) + 1.933(\log_{10}[NOx])^2 - 0.326(\log_{10}[NOx])^3\}$$

The concentrations denoted by the square brackets are in ppb.

This expression is plotted in Figure 9.7 and has been proposed as a useful and quick guide to the behaviour of the NOx system in urban atmospheres.

Figure 9.7 Relationship Between Urban Hourly Mean NO_2 Measurements and NOx

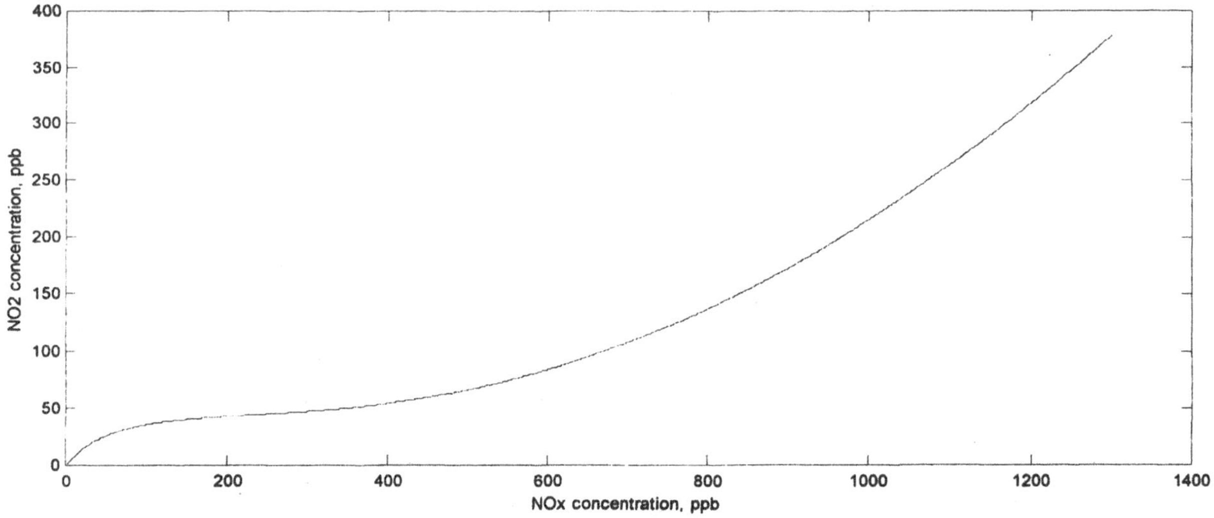

Alternatively, a theoretical relationship can be adopted, using representative values for the ozone concentration in background air and solar radiation, which determines the photo-oxidation rate of NO_2. Such photochemical models sometimes form part of other models. There is no correct way to predict NO_2 concentrations, and model validation is important. Further examples of NOx and NO_2 relationships have been published in more detailed pollutant specific guidance.†

* Derwent R G & Middleton D R, 1996, 'An Empirical Function for the Ratio NO_2:NOx', *Clean Air*, 26 (3/4) 57–60

† *Review and Assessment: Pollutant Specific Guidance*, LAQM.TG4(00), DETR, London, 2000

Long Range Dispersion Model Evaluation – ETEX

Since the Chernobyl accident in April 1986 the paucity of controlled and well-documented experimental data, suitable for the evaluation of long range atmospheric dispersion models has been evident. To explore the performance of models over a wider range, a long range European-scale Tracer EXperiment named 'ETEX' was carried out in October 1994. The ETEX experiment was organized as a continuation of the joint Atmospheric Transport Model Evaluation Study (ATMES) initiated and sponsored by the EU, the World Meteorological Organization (WMO) and the International Atomic Energy Agency (IAEA). The objectives of ETEX were:

- to conduct a long range atmospheric tracer experiment;
- to test the capability of institutes involved in emergency response to produce predictions of the cloud evolution in real time;
- to evaluate the validity of their predictions; and
- to assemble a database for evaluating long range atmospheric dispersion models in general.

During ETEX a tracer gas cloud was monitored as it dispersed over Europe over several days. After several hundreds of permanent sampling stations were disseminated throughout the central parts of Europe the time and place of the release was announced only shortly (hours) before the release.

The first of two tracer releases, ETEX-1, started on 23 October 1994 at 16:00 UTC and ended 12 hrs later. A total of 340 kg of the tracer gas perfluoro-methyl-cyclohexane (PMCH) was released at a constant release rate of 7.95 g/s from the top of a stack (8 m tall) located 35 km west of Rennes in Brittany, France. Advantages of the PMCH-tracer are that it is non-toxic, it has little or no wet and dry deposition, and, maybe most importantly, its atmospheric background concentration is extremely low, about 4×10^{-15} parts by volume. Moreover, PMCH is detectable via gas chromatography down to very low concentrations.

During the ETEX-1 release, the Danish National Environmental Research Institute (NERI) carried out real time measurements of the tracer from a well exposed site (the harbour) at Risø, Roskilde, Denmark; more than 1200 km from the release point. The tracer concentration was analysed at hourly intervals by gas chromatography. Air samples were also collected both in saran sample bags and by the official ETEX sample tubes provided by the Ispra Joint Research Centre. The saran bags were used to collect 1 hr mean air samples continuously throughout the campaigns whereas the ETEX sample tubes were used to take 6 min mean air samples once every hour.

The tracer plume was first observed at Risø 44 hr after the start of the release in France and the cloud lasted above the threshold detection level for 20 hr. The tracer concentration varied considerably during the passage. Two peaks were observed, with maxima of about 13 and 5 times the background concentration. This concentration variation was observed both in the data set resulting from the saran bags and in the data set of the sample tubes. DREAM (the Danish Rimpuff and Eulerian Accidental Release Model) is a high-resolution three-dimensional tracer model, developed for studying short and large scale atmospheric transportation, dispersion, and deposition (wet and dry) of radioactive air pollution caused by a single but strong source, as for example, the Chernobyl accident. The model is based on a combination of a Lagrangian short scale puff model and an Eulerian long range transport model. The Lagrangian model is used in the area near the source to describe the transportation, dispersion and deposition in the initial phase of the release and the Eulerian model is used for long range transport, dispersion, and deposition calculations in the model domain over the whole of Europe. The model was run and validated against the ETEX releases. Figure 9.8 illustrates the DREAM model results of the situation 12 hr and 48 hr after the start of the release. The square indicates the area where the Lagrangian model operated. The Eulerian model applied over the whole model domain. The NAME model introduced previously also showed a good performance in the ETEX study of in terms of the correlation between observed spread and model predictions.

Figure 9.8 Model Simulations of the Surface Concentrations of the Tracer (PMCH) Released from Britanny, France, During the European Tracer EXperiment (ETEX-1)

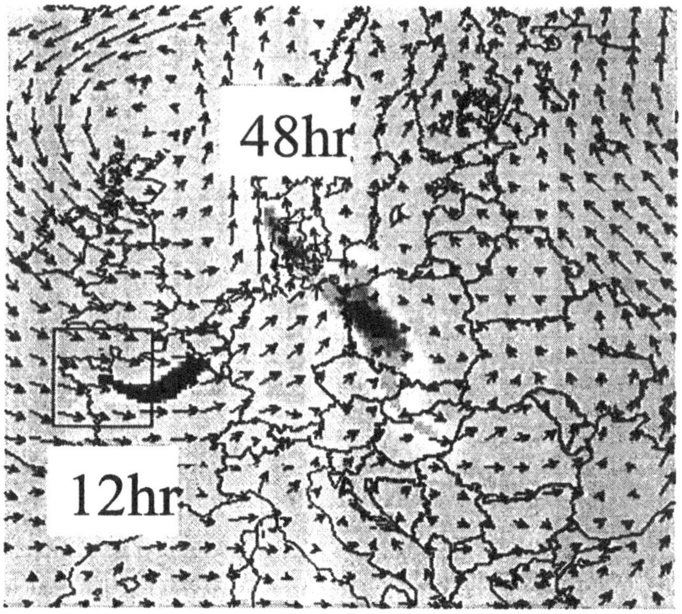

Overall, the ETEX study demonstrated the feasibility of conducting a continental scale tracer experiment across Europe using the perfluorocarbon tracer technique. Differences between observations and calculations of 3 to 6 hr in arrival time and a factor of 3 in maximum surface concentrations were viewed as the best achievable with current models – well within the range suggested in Table 9.1. Looking at the results predicted for 48 hr after release start, it was evident that the forecast meteorological fields were in many cases sufficiently accurate to allow models to predict the cloud position and extension. A second ETEX release showed larger discrepancies between observations and model results where all models significantly over-predicted surface concentrations after 12 hr from the start of the release. No clear explanation was identified for this result. Further investigations are needed to determine the quality of predictions under complex meteorological conditions, and to quantify the uncertainty of models for emergency management. Not surprisingly, the meteorological fields produced by the global circulation models give rise to concentration values that are less accurate than those obtained using results from finer scale meteorological models.

Recommendations resulting from the ETEX project included the following points:

- The dispersion models require meteorological parameters such as boundary layer height and vertical diffusion coefficients, from the output of the limited area meteorological models. Discrepancies between tracer concentration measurements and results from dispersion modelling indicate that more work has to be done to derive these parameters. Furthermore, the processing of meteorological data from flow models, as input to dispersion models needs to be improved. Particular attention in future work should be given to the parameterization of convective processes, low wind conditions and precipitation.
- Methods that take into account all of the available information during an emergency should be investigated. Particular attention should be given to the assimilation of measured concentrations.
- Knowledge of the vertical structure of the dispersed plume is essential for understanding the dispersion process. Measurements of concentration distribution in the vertical are very important.

Conclusion

A number of studies have been carried out to examine the reliability of predicted concentrations required for evaluations in local air quality management. The results of these studies confirm long held views that:

- models are more reliable for estimating longer time-averaged concentrations than for estimating short term concentrations at specific locations;
- models are reasonably reliable in estimating the magnitude of highest concentrations occurring sometime, somewhere within an area – however, errors in highest estimated concentrations often exceed the factor-of-two accuracy that has long been quoted for these models; and
- estimates of concentrations that occur at a specific time and site, are poorly correlated with actually observed concentrations and are less reliable.

With any modelling system, the accuracy of the predictions depends on the quality of the emissions database, and studies have found that this input must be comprehensive and of high quality. Care should also be taken in interpreting the statistics for paired time series of monitored and predicted values. Models typically consider a steady state and assume that all pollutants emitted during an hour will arrive at the receptors during the same hour. Therefore, the model may predict well the levels of concentration, but with a time variation. In the modelling of industrial emissions, the estimations of short term peak concentrations are always subject to more uncertainty than the predicted annual mean values. For example, the maximum hour in any year is a very variable quantity, depending as it does on fluctuating local meteorological conditions, whereas the estimated annual average concentration tends to be a more reliable output, as short term weather variations are averaged out.

By performing periodic trials and comparing model results one is able to reduce the discrepancies in model predictions and thereby to improve the confidence of decision-makers in model support for air quality management and for use during emergencies.

CHAPTER 10
MODELLING GUIDELINES

This final chapter aims to draw together principles from previous chapters and relate them to guidelines for modelling. The management of urban air quality is a complicated issue, requiring:

- a sufficiently detailed set of data for the urban area being studied (topography, building arrangements, sources, wind etc);
- a methodology for constructing scenarios;
- models for predicting the dispersion of the air pollutant over the area; and
- a decision support methodology able to integrate the various modules and present appropriate choices for decision-making.

If information is needed about direct health effects (short term or long term), methodologies need to be added for the prediction of such effects, based on the output of the dispersion model. Population density and location data are also needed. Overall, a complex set of data, models and support tools needs to be integrated to provide sufficient information for managing air quality in a coherent way.

Model selection requires matching the important characteristics of the site and requirements of the evaluation with the capabilities of the model and normally demands expertise. As a general principle, modelling should start with the simplest form possible, moving to more complex approaches only where the necessity and value of such modelling can be demonstrated. At the basic level, a crude mass balance can indicate whether a new source is likely to pose a problem. Alternatively, a simple screening model can give a realistic estimate of the magnitude of the impacts of a source. Situations with multiple sources or with varying terrain may require a more sophisticated effort, involving site-specific data collection and more complex models. There are cases where more than one model may be required. For example, modelling around a valley requires the use of one model for the valley floor with flat terrain and another for the high ground around the valley.

Any model needs to be solved and applied with representative field data (either directly measured or indirectly estimated) to calculate pollutant concentrations at various distances and for different meteorological conditions. Typical data required to 'operate' a Gaussian plume model for an elevated point source were developed in previous chapters and are summarized in Table 10.1

Table 10.1 Data Requirements for Dispersion Models

Parameters that may be directly measured	Parameters that may be estimated indirectly
Physical height of the stack	Dispersion coefficients
Stack diameter	Plume rise including stack downwash if any
Average stack velocity	Atmospheric stability
Gas exit temperature	Wind velocity at the top of the stack
Average ambient temperature	Mixing height (using temperature data)
Design source strength	
Wind velocity and directions at ground level	
Cloud cover and solar insolation	
Mixing height	

Resource Requirements for Dispersion Studies

Air quality modelling can be a complex task and the objectives need to be clear. The costs of a study depend on the complexity of the situation and the level of detail required, but often costs can be at the lower end of this scale. Information on screening models is generally readily available. The costs for acquiring the model, some training and the actual study should be less than £5000, with some models being available free of charge. Moreover, people can rapidly acquire skill with the screening models. On the other hand, where refined modelling is required, then the necessary skill level increases significantly.

Air quality monitoring and model validation can have significant costs. In the US where modelling is a statutory requirement for new source permits, costs for power plants have ranged from $100,000 to $2 million. The lower end of this range corresponds to the case of readily available meteorological data and flat terrain in a rural area. The high end of the range includes ambient air quality monitoring costs and, in some cases, the cost of demonstrating the inappropriateness of a model approved by the regulatory agency or validating a model that is not approved by the agency.

Moving up the scale leads to a point source dispersion model that estimates the ground-level concentrations of pollutants of interest at some distance (typically from hundreds of metres to tens of kilometres) from a point source. More complicated models allow the examination of multiple sources, including area sources (ie non-point). For an area containing a number of point and non-point sources, an air quality model could be constructed that includes all of the sources in the area. In practice, such models are rare because of the costs of development and the data required to make the model a realistic tool. Costs associated with an example described in Chapter 10 used a workstation with standard installation £14,850 plus a colour printer with Postscript option £1,870. Software (Indic AirViro Emission database and dispersion modules) installation and configuration with a single user licence and on-site training for 1 week cost about £51,500 exclusive of VAT. Operator time is an additional cost.

A Practical Approach to Dispersion Modelling

When a user faces a new modelling application it is necessary to plan for the process. Given any modelling scenario, the user must analyse the data available and identify basic elements for the specific problem. These categories are typically based on the considerations for model control options, source data, receptor locations, meteorological conditions, and model output options.

Any modeller should start with a clear picture of the modelling scenario and write down a brief description of the problem. Then list the basic modelling options that are applicable to the problem. Model control options to consider include the following:

- Should concentration or deposition be calculated?
- Is the area to be modelled rural or urban?
- Does the modelling application need to meet the demands of a regulatory body?
- What pollutant or pollutants are involved?
- Based on the regulations applicable for the significant pollutants or the potential for 'harm', what averaging period(s) is/are of concern?
- Is the terrain flat or elevated?
- Are any of the pollutants reactive?

Source characteristics undoubtedly play a major role in air quality modelling. Without sources of pollutant emissions, it would not be necessary to perform any air quality analysis. While source analysis is very important, it can also be the most complicated part of the modelling, as the examples described in Chapter 9 illustrated. In order to set up a successful modelling run, the users should consider both foreground and background sources of pollutant emissions. Foreground sources are the primary sources for which the modelling is being done, perhaps as part of an application for an authorization or other regulatory purpose.

In the real world, background source information is important. It can sometimes be a crucial factor, especially for a source located within or near a sensitive or already polluted area. For example, a process that could readily obtain an authorization in one location may find it almost impossible to do so in another area where the background air quality resulting from other sources is already high. Modellers can often obtain background source information from statutory registers such as those under provisions of the Environmental Protection Act 1990 and this was illustrated in the practical case in Chapter 9. Background air quality data can come from national air quality networks or local authorities as discussed in earlier chapters.

For both foreground and background sources, and for all types of pollutants being considered, the modeller should determine the following points:

- The type of each source – is it a point source (such as a stack or isolated vent), an area source (for ponds, landfills, or storage piles), a volume source (for example, conveyors, structures with multiple vents, etc) or a line source (such as a motorway)?
- The location of the source – the coordinates for each source will be needed, relative to a common coordinate system. For areas with elevated terrain, the elevation (z-coordinate) of each source is needed.
- The emission rates of each pollutant must be determined for both nearby and background sources using approaches identified earlier.

Other source parameters depend on the type of source:

- For a point source, the effluent temperature, fluid velocity and release point diameter.
- For an area source, the width of the source.
- For a volume source, the initial height and width of the source.
- For a line source, the width, length, initial mixing zone and characteristics such as queuing.

To determine compliance with appropriate regulatory requirements such as compliance with air quality standards and the determination of BAT, it is often necessary to look at the combined impact of a group of sources. Additional optional considerations for source input data include the whether stack releases are influenced by nearby buildings. If so, the modeller will need to determine the effective width and height for each applicable building and carry out an appropriate analysis. It is also necessary to consider whether emissions vary with time, for example, by season or time of day or process activity. For modelling particulate matter, it may be necessary to determine the appropriate settling, removal and deposition parameters. These include the settling velocity categories, the mass fraction of material and the reflection coefficient for each category.

Receptors

A fundamental purpose of most dispersion modelling applications is to determine the maximum impact from the source or sources being modelled. To make this determination, the model must calculate impacts at a large number of receptor locations across the modelling domain. In addition, the user may want to determine the air quality impact at specific locations of interest, such as a nearby site of special scientific interest (SSSI), school, hospital or some other sensitive area. The user must therefore determine the number, type and location of all receptors to be modelled. Initial screening modelling may help to identify the areas of expected maximum impacts, especially for foreground sources. This knowledge is used in designing the receptor grids. Polar and Cartesian systems are two types of receptor networks commonly used in refined air dispersion models. A polar network (distance, direction) consists of concentric rings and radial arms extending outward from a centre point, usually the source being modelled as shown in Figure 10.1. Receptors are located where the concentric rings and radial arms intersect. Care should be exercised in using a polar network to identify maximum estimated pollutant concentrations owing to the inherent problem of increased longitudinal spacing of adjacent receptors as their distance along neighbouring radial arms increases. For example, while receptors on individual radials, may be uniformly spaced at a distance of 1 km apart, at greater distances from the proposed source, the longitudinal distance between the receptors on neighbouring radials may be several km. Consequently, finding the maximum source impact can be difficult in view of the increasing distance between receptors at the greater radial distances. For this reason, a polar network is less suitable for anything other than initial screening.

Figure 10.1 A Polar Network Has Increased Spacing Between Receptors Along Radii

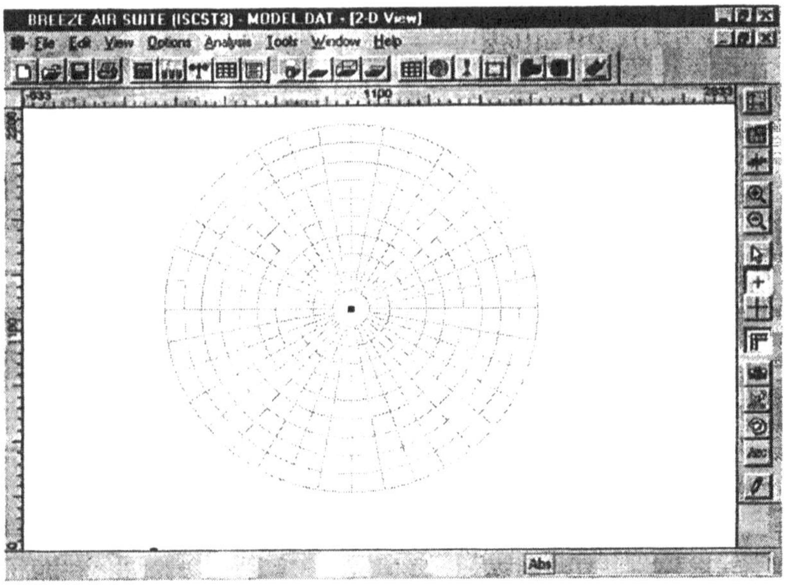

A Cartesian network (also referred to as a rectangular or x, y network) was shown in Figure 7.2. It consists of north-south and east-west oriented lines forming a rectangular grid with receptors located at each intersection point. In most refined air quality analyses, a Cartesian grid with from 300 to 400 receptors (where the distance from the source to the farthest receptor is 10 km) is often adequate to identify areas of maximum concentration. However, the total number of receptors will depend on the specific study.

Questions about other receptor characteristics include:

- How many grids of receptors are needed?
- How many discrete receptor locations are needed?
- Should receptors be placed at the site boundary?
- Is elevated terrain involved?
- Are flagpole receptor heights (above ground level) involved?

Elevated terrain is terrain that is above the stack base elevation and below the stack top elevation. In order to model the effects of elevated terrain with appropriate models, terrain elevations must be defined for all receptor locations. Flagpole receptors refer to the height of the receptor above local ground level. By default, models often assume ground level receptors and flat terrain.

In order to locate the maximum impact, multiple model runs start with a relatively coarse receptor grid (for example, 1 or 2 km spacing) and continue to a relatively fine receptor grid (for example, 100 m). The fine receptor grid serves to focus on the area(s) of higher estimated pollutant concentrations identified by the coarse grid model runs. Some air quality models allow discrete receptors at user-specified locations to be investigated. Receptor sites should be selected on a case-by-case basis, taking into consideration the topography etc. For example, receptors may be located at:

- the fence-line of a site under investigation;
- the boundary of a nearby sensitive area;
- the location(s) of ambient air monitoring sites; and
- locations where potentially high ambient air concentrations may occur.

In general, modelling receptors should be placed at ground level, but may be placed at elevated locations, such as on building rooftops if appropriate.

UK Guidelines

Guidelines on best practice in the use of mathematical models for atmospheric dispersion have been published by the Royal Meteorological Society in association with the then Department of the Environment.* Ten essential characteristics were highlighted:

1. Set out the situation being modelled, the objectives and how modelling aims to achieve them.
2. Demonstrate that the modelling procedure is fit for purpose.
3. Give full details of the model and its translation into a valid computer option.
4. Show how data requirements of a model are satisfied and deficiencies addressed.
5. Ensure the outcomes are successfully communicated to interested parties.
6. Express the findings quantitatively.
7. Demonstrate how the results vary with variables whose values are questionable.
8. Address the inherent variability in dispersion phenomena and in model variables.
9. Ensure that a model used has been validated to confirm its suitability for the task.
10. Ensure an audit trail for the example/exercise.

Guidelines have been published† to help local authorities to select and use models for the review and assessment of air quality. Guidelines were also issued to help applicants seeking authorizations under the Environmental Protection Act 1990.‡ Typically a tiered approach introduced in Chapter 5 serves to refine the estimate of the impact of a stationary source releasing a contaminant into the atmosphere. This approach minimizes the time and costs of securing regulatory approval. The stages start with a simple screening model such as workbook methods, R91 etc. If this shows that concentrations are only a small fraction of a relevant limit concentration, then the project may be regarded as acceptable without further modelling. The fraction of the standard should be based on the accuracy of the concentration estimates together with a safety factor. If simple screening suggests an unacceptable level, then the next level of model is appropriate. This could be a computer model such as SCREEN or ISC. Evidence of a problem with intermediate modelling suggests application of an advanced model such as AERMOD or ADMS. A similar sequence is found in US guidelines.

---

* *Atmospheric Dispersion Modelling: Guidelines on the Justification of Choice and Use of Models, and the Communication and Reporting of Results*, Royal Meteorological Society, 104 Oxford Road, Reading.

† *Selection and Use of Dispersion Models*, LAQM.TG3(00), Department of the Environment, Transport & the Regions.

‡ *Released Substances and Thir Dispersion in the Environment: Guidance for Applicants for Process Authorisation Under Integrated Pollution Control*, The Environment Agency.

US Guidelines

In the US, there are guidelines for evaluating the air quality impact of sources following the requirements of the Clean Air Act, and for new major or minor sources or modifications subject to new source review regulations. These guidelines follow the three-stage approach introduced in Chapter 5:

> **Stage 1:** Apply a simple screening procedure to determine if either the source clearly poses no air quality problem or the potential for an air quality problem exists.
>
> **Stage 2:** If the simplified screening results indicate a potential threat to air quality, further analysis is warranted and detailed screening (basic modelling) procedures should be applied.
>
> **Stage 3:** If detailed screening or other factors so indicate, make a more refined analysis.

Detailed modelling may be required, for example when:

- a more accurate estimate of the concentrations is needed (such as when the results of the Stage 2 analysis indicate a potential air quality problem);
- the source configuration is complex;
- emission rates are highly variable; and
- pollutant dispersion is significantly affected by nearby terrain features or large bodies of water.

In most of these situations computer based dispersion models can be of considerable help in estimating air quality impact. However, be aware that an air quality model applied improperly can lead to serious misjudgements regarding the source impact.

European Union Initiatives

The European Environment Agency (EEA) is an institution founded by the EU and located in Copenhagen, Denmark. The agency has the task of providing objective, reliable and comparable information at European level. This is to enable member states to take the requisite measures to protect the environment, to assess the results of such measures and to ensure that the public is properly informed about the state of the environment. To that end, the agency should provide the necessary technical and scientific support.

The central office of the EEA is not large and many projects are carried out by delegation to so-called European topic centres. One of these is the European Topic Centre on Air Quality, which is the most relevant to dispersion modelling. It is a consortium led by the Dutch RIVM with participation from NILU (Norwegian Institute of Air Research), DNMI (Norwegian Met Institute), and NOA (National Observatory of Athens). The topic centre has responsibility for several tasks including harmonization in the use of models for ambient air quality and pollution dispersion/transport. This work includes the development of toolkits for model evaluation in co-operation with the 'Initiative within Atmospheric Dispersion Modelling for Regulatory Purposes'. It also includes the establishment of a catalogue of dispersion models on the Internet.

When the catalogue was first launched in spring 1997, it included descriptions of around 30 models but is expected to expand as more modellers provide information to the system. The rationale for setting up an evaluation protocol was the belief that the traditional regulatory models in use today no longer represent the state of the art, and do not enable satisfactory assessments. An evaluation protocol will serve to show that more advanced models provide a better performance.

Quality control of improved regulatory atmospheric dispersion models can be achieved when it is possible to answer the following questions:

- How well do the models perform with well-defined meteorology ie, using as input all the available information that can be obtained from the above experimental data?
- How well do the models perform with degraded input, as compared to research quality input?

For any real industrial site one cannot expect to have all the required data. Some boundary-layer parameters will be obtainable only from the closest meteorological station. As the average spacing between these stations for European countries is in the order of 100 km, there is an expectation that models should be capable of producing reasonable results with approximated meteorological input. It is evident from discussion in previous chapters that up-to-date dispersion models, which have a better representation of atmospheric physics than earlier models, require as meteorological input some fundamental atmospheric boundary-layer parameters that are not directly measured.

Conclusion

This chapter and previous ones have indicated the background to modelling, a range of models available and the different levels of sophistication. The choice of model depends on the specific application, resource availability and expertise, but always the approach should start with the simplest appropriate model.

A starting point is to clarify what questions the modelling is required to answer. Usually, the first step in a modelling analysis is to estimate the ambient impacts caused by a source. This preliminary analysis establishes the impact area for each pollutant emitted in significant amounts, and for each averaging period. The largest impact area for each pollutant should be the impact area used in the full impact analysis. The emissions from all of the point, area and line sources at the proposed site are then modelled to estimate the area of significant impact for each pollutant. Where no significant concentrations of a pollutant are predicted at any location outside the boundary of the proposed source, that pollutant may be eliminated from further analysis. This preliminary analysis further indicates that monitoring data may be required for certain pollutants. Before undertaking a site-specific monitoring programme, the availability of existing data that are representative of air quality in the area is explored. National network sites should provide representative data.

After the impact area has been determined, the next step is to develop the required emissions inventories. These inventories provide source data that will be used to perform the dispersion modelling. Statutory registers from regulators should list all stationary sources in the area. As a conservative approach, emissions input data reflecting the maximum allowable emissions rate of each nearby source could be used in the modelling analysis. Air quality impacts caused by building downwash must be considered where appropriate. The necessary building dimension data are gathered for the inventory. The requisite input data (for example, source location, building dimensions, stack parameters, emissions factors and other emissions data) for the inventories are extracted from the registers.

Using input data from emissions inventories, the next step is to model existing source impacts using the chosen model. Estimation of total ground-level concentrations typically starts with a coarse receptor grid of 1 km grid spacing to identify the area(s) of high impact caused by the combined impact from the proposed new source and nearby sources. From the coarse grid run, the area of highest estimated concentrations is identified. In order to determine the highest total concentrations, a second model run is carried out for this area using a 100 m receptor fine-grid. The appropriate concentrations from the fine-grid run are added to the background concentrations to establish the total estimated concentrations for comparison against an ambient standard.

It has been suggested that there are more than 100 000 sources in Europe alone that may need assessment by site operators and regulators of the potential to cause adverse impact on air quality. Numerical modelling is a powerful and cost effective tool for air quality management and for emergency planning. As legal and social pressures for environmental improvements grow, so the role of models is likely to increase. The application of models will be facilitated by developments in emission inventory and air quality data, as well as in the performance of models themselves. Models need to be accessible, and the hierarchical approach encourages this, with the most advanced models being used only when simpler ones suggest a possible problem. In the end, though, it must not be forgotten that a model is a simplified description of a complex system. So long as it is remembered that models do not necessarily represent reality, they are still useful tools for air quality management.

APPENDIX 1
AIR QUALITY STANDARDS

The UK standards and those from other countries provide benchmarks for acceptable air quality against which model predictions (plus existing ambient air quality) need to be compared.

Table A1 The Air Quality Limit Values Regulations 2001, Statutory Instrument 2315

	Averaging period	Limit value	Margin of tolerance	Target date
Sulphur dioxide				
Limit value for the protection of human health	1 hr	350 μg.m^{-3}, not to be exceeded more than 24 times a calendar year	120 μg.m^{-3} on 19.7.2001, reducing yearly by equal annual amounts to reach 0 μg/m^3 by 1.1.2005	1.1.2005
Limit value for the protection of human health	24 hr	125 μg.m^{-3}, not to be exceeded more than 3 times a calendar year	None	1.1.2005
Limit value for the protection of ecosystems	Calendar year and winter (1.10 to 31.3)	20 μg.m^{-3}	None	19.7.2001
Alert threshold	3 consecutive hr*	500 μg.m^{-3}		
Nitrogen dioxide and oxides of nitrogen				
Limit value for the protection of human health	1 hr	200 μg.m^{-3} NO_2, not to be exceeded more than 18 times a calendar year	90 μg.m^{-3} on 19.7.2001, reducing yearly by equal annual amounts to reach 0 μg.m^{-3} by 1.1.2010	1.1.2010
Limit value for the protection of human health	Calendar year	40 μg.m^{-3} NO_2	18 μg.m^{-3} on 19.7.2001, reducing yearly by equal annual amounts to reach 0 μg/m^3 by 1.1.2010	1.1.2010
Limit value for the protection of vegetation	Calendar year	30 μg.m^{-3} NO_x	None	19.7.2001
Alert threshold	3 consecutive hr*	400 μg.m^{-3}		
Particulate matter (PM_{10})				
Limit value for the protection of human health	24 hr	50 μg.m^{-3} PM_{10}, not to be exceeded more than 35 times a calendar year	20 μg.m^{-3} on 19.7.2001, reducing yearly by equal annual amounts to reach 0 μg.m^{-3} by 1.1.2005	1.1.2005
Limit value for the protection of human health	Calendar year	40 μg.m^{-3} PM_{10}	6.4 μg.m^{-3} on 19.7.2001, reducing yearly by equal annual amounts to reach 0 μg.m^{-3} by 1.1.2005	1.1.2005
Lead				
Limit value for the protection of human health	Calendar year	0.5 μg.m^{-3}	0.4 μg.m^3 on 19.7.2001, reducing yearly by equal annual amounts to reach 0 μg.m^{-3} by 1.1.2005	1.1.2005

*Measured over the period at locations representative of air quality over at least 100 km^2 or an entire zone, whichever is the smaller

The figures for margins of tolerance are calculated from those given in Annexes I to IV to Directive 99/30/EC.

These Regulations that apply to England are made in the implementation of Council Directive 96/62/EC on ambient air quality assessment and management, and Council Directive 99/30/EC relating to limit values for sulphur dioxide, nitrogen dioxide and oxides of nitrogen, particulate matter and lead in ambient air. For more detail of interpretation the regulations should be consulted.

An EPAQS guideline for ozone is

Ozone 50ppb (100 $\mu g.m^{-3}$) as a 97th percentile of running eight-hr means.

It is not specified as a national objective because reaching it does not depend on local action alone.

In September 2001, acting on recommendations from the Department of Health's Committee on the Medical Effects of Air Pollution (COMEAP), the Department of the Environment, Food and Rural Affairs announced new standards for air pollution. The new targets included for the first time, one for polycyclic aromatic hydrocarbons (PAH).

Table A2 The UK Air Quality Strategy's Present and Proposed Objectives

Substance	Present Objective	Proposed Objective
Particulate matter (as PM_{10})	24 hr mean of 50 $\mu g.m^{-3}$ not exceeded more than 35 times a year. Annual mean of 40 $\mu g.m^{-3}$. Both to be met by end of 2004. Applied across the whole UK.	UK (excluding London & Scotland) 24 hr mean of 50 $\mu g.m^{-3}$ not exceeded more than 7 times a year. Annual mean of 20$\mu g.m^{-3}$. Both to be met by end of 2010. London 24 hr mean of 50$\mu g.m^{-3}$ not exceeded more than 10–14 times a year. Annual mean of 23 –25 $\mu g.m^{-3}$. Both to be met by end of 2010. A provisional 20 $\mu g.m^{-3}$ annual mean objective after 2010 was also proposed for London, with the aim of achieving it by 2015 where cost effective and proportionate local action can be identified. Scotland 24 hr mean of 50 $\mu g.m^{-3}$ not exceeded more than 7 times a year. Annual mean of 18 $\mu g.m^{-3}$. Both to be met by end of 2010.
Benzene	Running annual mean of 16.25 $\mu g.m^{-3}$ (5 ppb) to be met by end of 2003.	Running annual mean of 3.25 $\mu g.m^{-3}$ (1 ppb) to be met by end of 2010.
Carbon monoxide	Running 8 hr mean of 11.6 $mg.m^{-3}$ (10 ppb) to be met by end of 2003.	Running 8 hr mean of 10 $mg.m^{-3}$ (8.6 ppb) to be met by end of 2003.
Polycyclic aromatic hydrocarbons (PAHs)		Annual mean of 0.25 $ng.m^{-3}$ to be met by end of 2010.

Table A3 US Air Quality Standards

Substance	Averaging time	NAAQS ($\mu g\ m^{-3}$) Primary	Secondary
Sulphur dioxide	3 hr	–	1300
	24 hr	365	–
	Annual	80	–
Inhalable particles (PM_{10})	24 hr	150	150
	Annual	50	50
Inhalable particles ($PM_{2.5}$)	24 hr	65	65
	Annual	15	15
Nitrogen dioxide	Annual	100	100
Carbon monoxide	1 hr	40 000	–
	8 hr	10 000	–
Lead	3 months (calendar quarter)	1.5	1.5
Ozone	1 hr	235	235
	8 hr	157	157

Table A4 WHO 1996 Air Quality Guidelines for Europe

Substance	Averaging period	Concentration ($\mu g\ m^{-3}$)
Sulphur dioxide: health	10 minutes	500
	24 hr	125
	1 yr	50
Sulphur dioxide: ecotoxic effects	annual and winter mean	10 –30 depending on type of vegetation
Nitrogen dioxide: health	1 hr	200
	1 yr	40
Nitrogen dioxide and nitric oxide: ecotoxic effects	1 yr	30
PM_{10}	24 hr	dose/response
	1 yr	dose/response
Lead	1 yr	0.5

Concentration Conversions

Concentrations of air pollutants may be expressed in different units, which can be confusing. Two common systems are mass per unit volume (often $\mu g\ m^{-3}$) or as a volume mixing ratio (such as parts per million, ppm or parts per billion, ppb). As the volume of a gas depends on pressure and temperature, concentrations expressed in the former units should be qualified by the appropriate conditions. The variability of molecular volume indicates the sensitivity to conditions. For example, a change in temperature from 0°C to 27°C causes a 10% change in molecular volume. Volume ratios relate the amount of pollutant to the air in which it is contained. As temperature and pressure influence both to the same extent, the expressed concentration is independent of these variables. Concentrations of particulate matter cannot be expressed as a volume mixing ratio.

Conversion between a concentration in mass per unit volume and volume ratio such as ppm is a function of temperature, pressure and molecular mass, and uses the expression:

$$\frac{\mu g m^{-3}}{1000} = ppb \times \frac{\text{molecular mass (g)}}{\text{molecular volume (litres)}}$$

where the molecular volume $= 22.41 \times \frac{T}{273} \times \frac{1013}{P}$

T = absolute temperature (K)
P = atmospheric pressure (mb)

$$ppm = \frac{\mu g.m^{-3}}{1000} \times \frac{\text{molecular volume (litres)}}{\text{molecular mass}}$$

Some conversion factors applicable for an atmospheric pressure of 1013 mb are given in Table A8.

Table A5 Conversion Factors for Some Important Air Pollutants

Pollutant	Molecular mass	Conversion factors: $\mu g\ m^{-3}$ to ppb		ppb to $\mu g\ m^{-3}$	
		0°C	20°C	0°C	20°C
Carbon monoxide	28	0.80	0.86	1.25	1.17
Nitrogen dioxide	46	0.49	0.52	2.05	1.91
Nitrogen monoxide	30	0.75	0.80	1.34	1.25
Ozone	48	0.47	0.50	2.14	2.00
Sulphur dioxide	64	0.35	0.38	2.86	2.70

APPENDIX 2
LAND USE CATEGORIES

Classifying the land use near sources of air pollution is necessary because in general, urban areas cause greater rates of dispersion than rural areas. This is due to increased turbulent mixing and buoyancy-induced mixing caused by the combination of greater surface roughness caused by more buildings and structures and greater amounts of heat released from concrete and similar surfaces. Two procedures that serve to determine whether the character of an area is predominantly urban or rural are based on land use typing and on population density. Both procedures require an evaluation of characteristics within a 3 km radius from a source. The land use typing method is based on the work of August Auer and is more commonly used because it is more directly related to the surface characteristics of the evaluated area that affect dispersion rates. The assessment is made by identifying the land use made within 3 km of the site under investigation. Urban dispersion coefficients are appropriate if over 50% of the area falls within land use categories I1, I2, C1, R2 and R3, otherwise rural coefficients should be used. Areas of water are excluded from the assessment.

Land use analysis aims to estimate the percentage of the area within the 3 km radius that is either urban or rural. The type that has the greatest percentage will be the representative type. The most difficult evaluation involves the residential types in the table. The degree of resolution between types of residential areas often cannot be determined without conducting a site area inspection and referring to maps and aerial photographs.

Table A6 Auer Land Use Categories

	Land use category	Use	Composition	Vegetation
If more than 50% of the area falls in these categories, use urban coefficients.	**I1**	Heavy industrial	Major chemical, steel and fabrication industry. Generally 3 to 5 storeys and flat roofs.	Grass and tree growth very sparse – less that 5% vegetation.
	I2	Light to moderate industry	Rail and truck depots, warehousing, industrial parks and minor fabrication. Generally 1 to 3 storey buildings with flat roofs.	Limited grass and probably no trees. Less than 5% vegetation.
	C1	Commercial	Office and apartment buildings, and hotels. Generally over 10 storeys with flat roofs.	Limited grass and trees. Less than 15% vegetation.
	R2	Compact residential	Single, with some multiple, family dwellings, closely spaced. Generally not more than 2 storeys, with pitched roofs, garages, but no drives.	Limited lawns and shade trees. Less than 30% vegetation.
	R3	Compact residential	Old dwellings in multiple occupation. Flat roof structures within 2 m; garages, no driveways.	Limited lawns, old established shade trees. Less than 35% vegetation.

If more than 50% of the area falls in these categories, rural coefficients are appropriate	**R1**	Common residential (normal easements)	Single family dwellings, normally one storey with pitched roofs. Dwellings have driveways.	Abundant lawn areas and light to moderate tree cover. Over 70% vegetation.
	R4	Rural Estate residential	Expansive family dwellings across several acres.	Abundant lawns and lightly wooded area offering over 80% vegetation.
	A1	Metropolitan natural	Major municipal or similar parks, golf courses, cemeteries, campuses, with occasional single storey structures.	Almost total grass cover with lightly wooded areas. Over 95% vegetation.
	A2	Agricultural rural		Local crops, 95% vegetation.
	A3	Undeveloped	Uncultivated wasteland	Mostly wild grasses and weeds, lightly wooded. Typically 90% vegetation.
	A4	Undeveloped rural		Heavily wooded with over 95% vegetation.
	A5	Water surfaces		Rivers and lakes.

The Auer land use typing method can be labour intensive to apply. A simplified technique referred to in the hazardous waste combustion screening approach contained in the Title 40 Code of Federal Regulations, Part 266 (40 CFR 266), Appendix IX, Section 6.0 can be used as a screening tool. If the land use designation is clear, that is, about 70% or more of the total land use is either urban or rural, then further refinement is not required.

APPENDIX 3
SELECTED MODEL SUMMARIES

This is not intended to be a comprehensive list, but merely identifies some models that are available in addition to those in the main text.

ARCHIE is the acronym for 'Automated Resource for Chemical Hazard Incident Evaluation'. The model uses simple methods to perform release rate, pool evaporation, neutral and dense gas dispersion, pool fire, jet fire, fireball, BLEVE and vapour cloud explosion calculations. No chemical database is included and so the user must provide chemical data. Little or no modelling experience is required.

CAR (Calculation of Air Pollution from Road Traffic) International* was developed by TNO in the Netherlands. It has been administered by the Netherlands National Institute of Public Health and Environmental Protection (RIVM) since its development and uses readily available input data. Input requirements include emission factors; number of cars; percentage of heavy duty vehicles; driving speed. Output quantities include annual average concentrations and percentiles (5 to 30 m from road axis).

DREAM (the Danish Rimpuff and Eulerian Accidental Release Model) is a high-resolution three-dimensional tracer model, developed for studying short and large scale atmospheric transport, dispersion, and deposition (wet and dry) of radioactive air pollution caused by a single but strong source, as, for example, the Chernobyl accident. The model is based on a combination of a Lagrangian short scale puff model and an Eulerian long range transport model.

EDMS – Emissions and Dispersion Modeling System. A US Federal Aviation Administration (FAA) and US EPA 'preferred guideline' model for use in civil airports and military bases. EDMS was designed to assess the air quality impacts of airport emission sources, especially aviation sources. It can also model other airport sources including power plant, fuel storage tanks and vehicular sources, all of which are not aviation specific.

GASTAR was developed by CERC to model the consequences of dense and neutrally buoyant releases for both land use and emergency response planning. The model uses Monin–Obukhov length or Pasquill–Gifford stability categories. Other essential information includes wind speed and bearing, air and surface temperatures, atmospheric pressure, relative humidity and roughness length. Release types supported are: puff, plume, transient, jet, aerosol, thermal or isothermal releases. Output options include concentration puff footprint, cloud dimensions, temperature, density, volume, mass and position. Output for GASTAR may be presented graphically within its integrated graphics screen.

LADM (Lagrangian Atmospheric Dispersion Model) is an air quality modelling system from CSIRO Division of Atmospheric Research in Australia.† It comprises a mesoscale model for predicting winds and turbulence (LADM-M), and a Lagrangian particle model (LADM-P), which uses the winds and turbulence to predict pollutant dispersion. LADM-M allows the input synoptic pressure gradient to vary in the vertical, but it assumes that this vertical profile is horizontally homogeneous. Thermally forced flows like sea breezes can be simulated and also topographically forced flows like

---

* TNO Institute of Environmental Sciences, Energy Research and Process Innovation AH Apeldoorn, the Netherlands.

† Physick, W L, Noonan J A, McGregor, J L, Hurley, P J, Abbs, D J and Manins, P C, 1994, *LADM: A Lagrangian Atmospheric Dispersion Model*, CSIRO Division of Atmospheric Research, Technical Report No. 24.

drainage winds. LADM-P transports and diffuses particles according to the predicted wind and turbulence fields from LADM-M. Particles can be released from sources at any height and can represent reactive or inert gases. At each timestep the position co-ordinates of the particles are computed in terms of the LADM-M grid. Concentrations are calculated by counting particles in the grid volumes. As well as point-source emissions, LADM-P is able to model emissions from area/volume sources and line sources. It can also be used to model photochemical formation and the dispersion of gases such as ozone (O_3) and nitrogen dioxide (NO_2).

The Oklahoma Dispersion Model* was developed to aid agricultural and natural resource managers. The model was designed for near-surface release of gases and small particulate matter (airborne solid and liquid particles with diameters less than 20 μm). Examples include pesticides released by surface or aerial spraying, odours associated with the land application of animal waste and smoke from open burning.

RAINS, the 'Regional Air Pollution Information and Simulation' model was developed by the International Institute for Applied Systems Analysis (IIASA) as a tool for the integrated assessment of alternative strategies to reduce acid deposition in Europe and Asia†. The RAINS 7.2 model describes the pathways of emissions of sulphur dioxide, nitrogen oxides and ammonia and explores their impacts on acidification and eutrophication.

RMP*COMP is an electronic tool used to perform the off-site consequence analysis required under the Risk Management Program rule published by the Environmental Protection Agency on July 20, 1996, which implements Section 112(r) of the Clean Air Act.

SLAB is a dispersion model for denser-than-air releases, developed by the Lawrence Livermore National Laboratory, with support from the US DOE, USAF Engineering and Services Center, and the American Petroleum Institute. The user must provide chemical data as no chemical database is included, and substantial modelling experience is required. SLAB predicts hazardous zone and the potential impacts of accidental releases and meets US EPA Risk Management Plan requirements. It can model continuous, finite duration and instantaneous releases from four types of sources: a ground-level evaporating pool, an elevated horizontal jet, a stack or elevated vertical jet and a ground-based instantaneous release. Various types of graphical output can be produced including footprint of the puff released to the atmosphere, contour plots and XY plots showing maximum concentration versus downwind distances. The SLAB program can be downloaded from the SCRAM Internet site free of charge.

TOXST is the acronym for 'Toxic Modelling System Short-term'. It estimates the dispersion of toxic air pollutants from point, area and volume sources at a complex industrial site, and allows the estimation of ambient concentration impacts for single and multiple pollutants from continuous and intermittent sources. In addition, the model estimates the average annual frequency by which user-specified concentration thresholds are expected to be exceeded at receptor sites around the modelled site.

TSCREEN is the acronym for 'Toxic Screening Model'. It performs release rate, pool evaporation, and natural and dense gas dispersion calculations. As simple methods are used, the model demands little or no modelling experience. It is primarily used for screening. Briggs plume-rise equations are employed for buoyant emissions.

* Carlson, J D, 'Agricultural and Fire Meteorologist, Biosystems and Agricultural Engineering', Oklahoma State University, Stillwater, OK 74078.

† Alcamo J, Shaw R, and Hordijk L (eds), 1990: *The RAINS Model of Acidification. Science and Strategies in Europe,* Dordrecht, Netherlands: Kluwer Academic Publishers.

APPENDIX 4
SELECTED SOURCES OF ADDITIONAL INFORMATION

Atmospheric Dispersion Modelling: Guidelines on the Justification of Choice and Use of Models, and the Communication and Reporting of Results, Royal Meteorological Society, 104 Oxford Road, Reading.

Beychock, M R, *Fundamentals of Stack Gas Dispersion*, Milton R Beychock, Irvine Ca 92612, ISBN 0964458802.

Design Manual for Roads and Bridges, Volume 11, Section 3, Department of the Environment, Transport and the Regions, May 1999.

Guideline on Air Quality Models (Revised), EPA-450/2-78-027R, (Appendix W of 40 CFR Part 51) July 1986 US Environmental Protection Agency, Office of Air and Radiation, Office of Air Quality Planning and Standards, Research Triangle Park, North Carolina 27711.

Middleton, D R (1997) *Manual on Modelling: A Guide for Local Authorities.* Meteorological Office Turbulence and Diffusion Note Number 241.

Released Substances and their Dispersion in the Environment: Guidance for Applicants for Process Authorisation under Integrated Pollution Control, HMSO, 1996, ISBN 0 702010 9.

Review and Assessment: Pollutant Specific Guidance, LAQM.TG4(00), Department of the Environment, Transport and the Regions, May 2000.

Screening Procedures for Estimating the Air Quality Impact of Stationary Sources. Revised EPA-454/R-92-019, US Environmental Protection Agency, Office of Air and Radiation, Office of Air Quality Planning and Standards, Research Triangle Park, North Carolina 27711.

Selection and Use of Dispersion Models, LAQM.TG3(00), Department of the Environment, Transport and the Regions, May 2000.

Schulze, R H, (1991) *Practical Guide to Atmospheric Dispersion Modeling*, Trinity Consultants, Inc, Dallas, Texas.

Turner, D B (1994) *Workbook of Atmospheric Dispersion Estimates*, 2nd edition, Lewis, ISBN 156670023X.

Zannetti, P (1990) *Air Pollution Modelling: Theories, Computational Methods and Available Software*, Van Nostrand Reinhold, New York.

APPENDIX 5
WEB ADDRESSES

All visited 19 September 2001

Aeolius	http://www.meto.govt.uk/
AIRS Executive database of air quality data from nations contributing to the GEMS/AIR programme sponsored by WHO	http://www.epa.gov/airs/aexec.html
BEE-line software	http://www.beeline-software.com/software.html
California Air Resources Board	http://www.arb.ca.gov/homepage.htm
Cambridge Environmental Research Consultants Ltd (ADMS and GASTAR)	http://www.cerc.co.uk
Catalogue of dispersion models (European Environment Agency)	http://aix.meng.auth.gr/database/index.html
CORINAIR 90 study of atmospheric emissions in Europe	http://www.aeat.co.uk/netcen/corinair/corinair.html
Databases and software for chemical emergency preparedness and prevention	http://www.epa.gov/swercepp/tools.html
European Environment Agency	http://www.eea.eu.int/
European Topic Centre on Air and Climate Change	http://etc-acc.eionet.eu.int/
GAIA model base – a database for models for air quality	http://www.ess.co.at/GAIA/models.html
Gaussian calculation	http://www.industrialhygiene.com/calc/model.html
Helpline for local authorities on modelling. Includes DMRB spreadsheet	http://www.stanger.co.uk/airqual/modelhlp/helpline.htm
Hysplit – description and how to access the model.	http://www.arl.noaa.gov/hysplit.html#INFO
International Institute of Environmental Sciences and Environmental Computing	http://www.envirocomp.org/
Lakes Environmental Software	http://www.lakes-environmental.com/
UK air quality management resource centre (University of the West of England)	http://www.uwe.ac.uk/aqm/centre/index.html
Model database	http://www.mi.uni-hamburg.de/cgi-bin/cost/models/database_order.cgi
Model and model harmonization	http://www.dmu.dk/forside.asp
National Society for Clean Air	http://www.nsca.org.uk
PARAMICS traffic model	http://www.epcc.ed.ac.uk/epcc-projects/PARAMICS/
Real Time Applications and Display System (READY): interactive dispersion modelling	http://www.arl.noaa.gov/ready.html
Regional Specialized Meteorological Center for Transport & Dispersion Model Products	http://www.arl.noaa.gov/research/ep/rsmc.html
Scientific Software Group	http://www.scisoftware.com/
Technical Guidance Note 3, Selection & Use of Dispersion Models	http://www.defra.gov.uk/environment/airquality/laqm.htm
Trinity Consultants Inc (BREEZE software)	http://www.breeze-software.com/
UK national air quality information archive	http://www.aeat.co.uk/netcen/airqual/welcome.html
UK Emission Factors Database	http://www.rsk.co.uk/ukefd/index.htm
US EPA Office of Air and Radiation	http://www.epa.gov/oar/
US EPA Support Center for Regulatory Air Models (SCRAM)	http://www.epa.gov/scram001/main.htm

APPENDIX 6
SOLUTIONS TO EXERCISES

Chapter 3

1. a) t_{tot}= 8760; t_a= 24; t_b= 1, so substituting these values gives:

$$v = \frac{\ln\left(\frac{8760}{1}\right)}{\ln\left(\frac{8760}{24}\right)} = 1.54$$

Then:

$$m_{gb} = 0.067\exp[0.5(1-1.54)(\ln 1.88)^2] = 0.06ppm$$

b) t_{tot}= 8760; t_a= 24; t_b= 8760, so substituting these values gives:

$$v = \frac{\ln\left(\frac{8760}{8760}\right)}{\ln\left(\frac{8760}{24}\right)} = 0$$

Then:

$$m_{gb} = 0.046\exp[0.5(1-0)(\ln 1.72)^2] = 0.053ppm$$

2. Table 3.7 gives the surface wind speeds in units of m.s^{-1}. Hence the first task is to convert the values in the question to the appropriate units. Having done that the stability categories are assigned as follows:

a) On 28 December, with only 1 hr of bright sunshine (ie slight incoming radiation) and a wind speed of 5.1 km.hr^{-1} (i.e. 1.4 m.s^{-1}), the stability is class B.

b) On 23 August, with 11.4 hr of bright sunshine (ie strong incoming radiation) and a wind speed of 11.9 km.hr^{-1} (i.e. 3.3 m.s^{-1}), the stability category is B once again.

c) On 27 September, with 3 hr of bright sunshine (ie moderate incoming solar radiation) and a wind speed of 28.8 km.hr^{-1} (i.e. 8 m.s^{-1}), the stability category is D.

3. Based on 1 hr averaging times for the air quality data, standard geometric deviations of 1.3 and 1.6 are between the values of 1.10 and 1.75 in the table. Hence moving to the top of the table, the ratio of 1 hr maximum concentration to mean concentration will be between 1.43 and 7.21. By interpolation, factors of about 2.63 and 5.38 are suggested. These were the multipliers used by the regulatory body.

4. From the dry adiabatic lapse rate, it is possible to predict the theoretical rate of change of temperature with height, and to compare this with the actual environmental lapse rate as follows:

Altitude (m)	Temperature (°C)	
	Environmental	Predicted by dry adiabatic lapse rate
0	12	12
100	10.8	11.024
200	9.6	10.048
300	8.4	9.072
400	7.2	8.096
500	6	7.12
600	6	6.144
700	6	5.168
800	6	4.192
900	6	3.216
1000	6	2.24
1100	6.976	1.264
1200	7.952	0.288
1300	8.928	-0.688
1400	9.904	-1.664
1500	10.88	-2.64

The change may be more clearly represented diagrammatically as shown in Figure A1.

Figure A1 Lapse Rate for Question 4

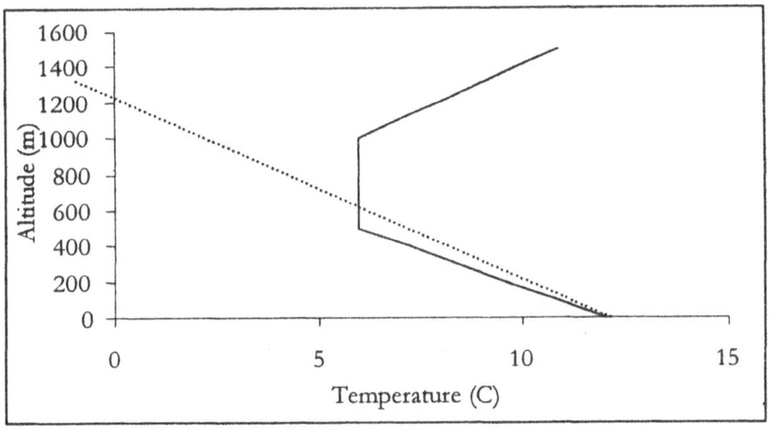

The lapse rate tends towards the superadiabatic below 500 m, but more important is the inversion above 1000 m that may trap pollution and lead to fumigation problems further afield.

5. The lapse rates for the situations in the question are represented in Figure A2.

Figure A2 Environmental Lapse Rates for Question 5

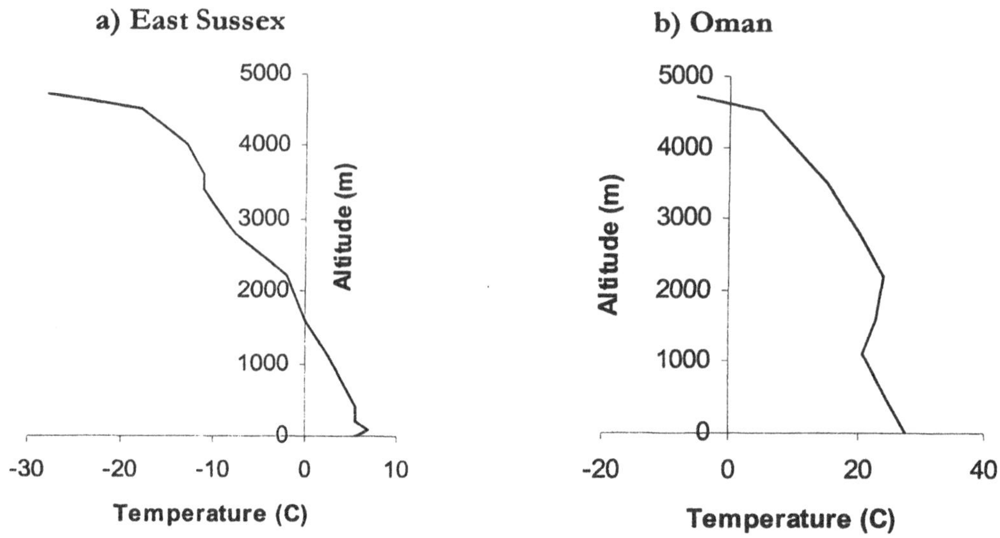

Differences in the temperatures are evident from these data, but the key factor is the rate of change compared with the dry adiabatic lapse rate. It is clear from a) that there is stable air near the surface characteristic of a radiation inversion typical of January night-time. Low level releases from traffic, for example, may be trapped, although releases from point sources penetrate the weak inversion.

By comparison, b) shows a substantial and elevated inversion with greater potential to trap pollution. The cause of this inversion is explained as follows. In the summer, Oman typically experiences a shallow south-westerly air flow of relatively cool air overlain by a hot north-westerly air mass from the arid Arabian interior. This leads to a marked temperature inversion along the coastal zone, completely suppressing convection and impeding the dispersion of pollutants.

Chapter 5

1. A worst case scenario assumes no plume rise. For stack 1, the downwind distance from the source = 2.8 + 0.4 = 3.2 km. The emission rate (Q) is given by:

$$Q = \frac{0.5 \times 1000}{24 \times 3600} = 5.8gs^{-1}$$

The weather conditions indicate Pasquill stability category D for slight to moderate insolation and a wind speed over 5 m s^{-1}. From the Figures 5.4 and 5.5, $\sigma_y = 180$ m and $\sigma_z = 74$ m.
The relevant dispersion equation is:

$$C = \frac{Q}{\pi u \sigma_y \sigma_z} \exp\left(-\frac{y^2}{2\sigma_y^2}\right)\left[\exp\left(-\frac{(H)^2}{2\sigma_z^2}\right)\right]$$

Therefore, substituting the appropriate values for stack 1 gives:

$$C_1 = \frac{5.8}{\pi \times 5.4 \times 180 \times 74} \exp\left(-\frac{480^2}{2 \times 180^2}\right)\left[\exp\left(-\frac{(74.5)^2}{2 \times 74^2}\right)\right] = 4.5 \times 10^{-7} gm^{-3}$$

For stack 2, the downwind distance is 3.2 km. The emission rate (Q) is given by:

$$Q = \frac{0.75 \times 1000}{24 \times 3600} = 8.7gs^{-1}$$

The dispersion coefficients have the same values as before, σ_y = 180 m and σ_z = 74 m, but y = 0. The relevant dispersion equation is:

$$C = \frac{Q}{\pi u \sigma_y \sigma_z}\left[\exp\left(-\frac{(H)^2}{2\sigma_z^2}\right)\right]$$

Therefore, substituting the appropriate values for stack 2 gives:

$$C_2 = \frac{8.7}{\pi \times 5.4 \times 180 \times 74}\left[\exp\left(-\frac{(100)^2}{2 \times 74^2}\right)\right] = 1.01 \times 10^{-5}\,gm^{-3}$$

For stack 3, the downwind distance from the source is 2.8 km. The emission rate (Q) is given by:

$$Q = \frac{0.25 \times 1000}{24 \times 3600} = 2.9gs^{-1}$$

From the graphs of dispersion coefficients at a distance of 2.8 km, σ_y = 160 m and σ_z = 66m. The value for the lateral distance is y = 320 m. Therefore, substituting the appropriate values in the same equation as used for stack 1 gives:

$$C_3 = \frac{2.9}{\pi \times 5.4 \times 160 \times 66}\exp\left(-\frac{320^2}{2 \times 160^2}\right)\left[\exp\left(-\frac{(102)^2}{2 \times 66^2}\right)\right] = 6.5 \times 10^{-7}\,gm^{-3}$$

The additional contribution to ground level concentration is the sum of that from the three sources:

$$C_{total} = C_1 + C_2 + C_3 = (4.5 \times 10^{-7}) + (1.1 \times 10^{-5}) + (6.5 \times 10^{-7}) = 12\mu gm^{-3}$$

Using the conversion factors in the appendix, this equates to 4.5 ppb. This is well below the levels likely to cause adverse health effects on its own, although the ultimate decision depends on the background air quality to which this additional burden must be added.

2. First find the buoyancy parameter from:

$$F = gvr^2\left(\frac{T_s - T}{T_s}\right)$$

$$F = 9.8 \times 20 \times 1^2\left(\frac{400-293}{400}\right) = 52.43\ m^4 s^{-3}$$

From Table 3.8, the stability class B applies for a sunny warm day, so the relevant equation is

$$\Delta h = \frac{1.6F^{\frac{1}{3}}(x_f)^{\frac{2}{3}}}{\upsilon}$$

Next:

$$x_f = 50F^{5/8} \text{ because } F < 55\ m^4\ s^{-3}$$

So:

$$x_f = 50 \times 52.43^{5/8} = 593.9$$

Then:

$$\Delta h = \frac{1.6 \times 52.43^{\frac{1}{3}}(593.9)^{\frac{2}{3}}}{3} = 133$$

The plume rise is therefore 133 m.

3. The buoyancy parameter, F, is first calculated:

$$F = gvr^2 \left(\frac{T_s - T}{T_s}\right) = 9.8 \text{ m s}^{-2} \text{ x } 8.8 \text{ m s}^{-1} \text{ x } (2 \text{ m})^2 \text{ x } \left(\frac{413-293}{413}\right)$$

Hence:

$$F = 100.2 \text{ m}^4 \text{ s}^{-3}$$

In the stable conditions with the atmospheric lapse rate of 2°C km^{-1} (ie 0.002 K m^{-1}).

$$s = \frac{g}{T}\left(\frac{dT}{dz}\right) = \frac{9.8ms^2}{293K}\left(0.002Km^1\right) = 0.00007 \text{ s}^{-2}$$

The final plume rise is then given by:

$$\Delta h = 2.4\left(\frac{F}{us}\right)^{\frac{1}{3}} = 2.4 \text{ x } 66.9 = 161 \text{ m}$$

Hence the effective stack height is:

$$H = h + \Delta h = 77 \text{ m} + 161 \text{ m} = 238 \text{ m.}$$

Chapter 6

1. On an overcast day, the stability class is D, the most common stability in the UK. To determine the source strength, the amount of sulphur burned in combustion of the fuel is:

$5 \text{ tonnes h}^{-1} \times 0.02 \times 1000 \text{ kg tonne}^{-1} = 100 \text{ kg h}^{-1}$.

The combustion of sulphur follows the equation:

	S	+	$O_2 \rightarrow$	SO_2
Molecular weights:	32		32	64

Hence every mass unit of sulphur burned produces two mass units of sulphur dioxide.

Source strength: $Q = 2 \times 100 \text{ kg h}^{-1} \times 1\text{h } 3600\text{s}^{-1} \times 1000\text{g kg}^{-1} = 55.6 \text{ g s}^{-1}$

Knowing that the effective stack height is 100 m and the stability category is D, the position of maximum ground level concentration (x_{max}) is read from Figure 6.3(a) as 3 km. The figure also gives a maximal value for $\chi u/Q$ as 8 x 10^{-6} m^{-2}. Hence:

$$x_{max} = (\chi u/Q)(Q/u) = 8 \text{ x } 10^{-6} \text{ x } 55.6$$
$$= 444 \text{ µg m}^{-3}.$$

Therefore the maximum concentration at ground level is 444 µg m^{-3} and occurs at 3 km distance from the stack.

2. The source strength is first calculated:

$$Q = \frac{10tonnes}{1h} \times 0.03 \times \frac{1000kg}{1tonne} \times \frac{1000g}{1kg} \times \frac{1h}{3600s} \times 2$$

=167 g sulphur dioxide each second.

Using Table 3.7 it is evident that a clear night with wind speeds of 4m s^{-1} suggests stability category E. If it is assumed that the power station is in a rural location, from Figure 6.3(a) it may be seen that the maximal ground level concentration will occur at about 13 km from the source. The figure also gives a value for the maximal value of (Xu/Q) of 1.7 x 10^{-6} m^{-2}. Consequently the maximal concentration is 1.7 x $10^{-6}m^{-2}$ x (167/4) = 71 µg m^{-3}.

So the ground level concentration of sulphur dioxide contributed by the plant is 71 µg m^{-3} and occurs 13 km from the source. Notice that this concentration is in addition to that resulting from all other sources surrounding the site. If all other emission characteristics are known, they may be estimated in a similar manner, and added together to give the total predicted contribution. This must be added to the ambient background to give the predicted air quality.

3. Figure 6.3 suggests that in rural terrain, the position of maximum concentration at ground level occurs at about 6 km and 1 km under D and B stability respectively. In urban terrain with B stability the maximum is 4 km from the source and 10 km under D stability. A realistic assumption is that a power plant would be built in a rural location, and so only this scenario will be considered.

Under B stability with the plume approaching the tower over rural terrain, the variation of concentration (μg m^{-3}) is represented in Figure A3. This graph relates to a distance of 1 km from the source at which the standard deviations are 160 m and 110 m respectively.

Figure A3 Concentration Profiles with height for a Source 1 km Distant in Rural Terrain

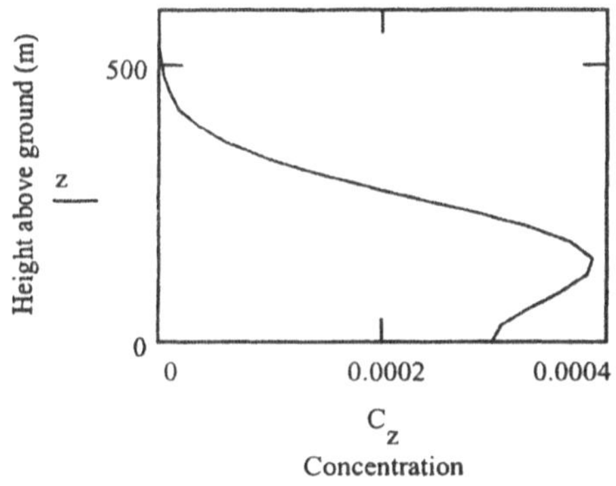

The maximum concentration is 390 μg m^{-3} and occurs at the plume height of 150 m. If the source were built in rural terrain some 6 km from the tower, the concentrations under D stability are as shown in Figure A4. In this case, the standard deviations are 300 and 100 respectively The concentration reaches a maximum at 150 m once more, but the dispersion over distance has reduced the concentration to 220 μg m^{-3}.

4. The graph in Figure A5 computed in MATHCAD® illustrate the point made in Chapter 3 that the greatest concentrations at ground level can occur during unstable conditions. In the example shown the greatest impact occurs in stability A for the stack height 300 m. The concentration scale is relative.

Figure A4 Concentration Profiles with height for a Source 6 km Distant in Rural Terrain

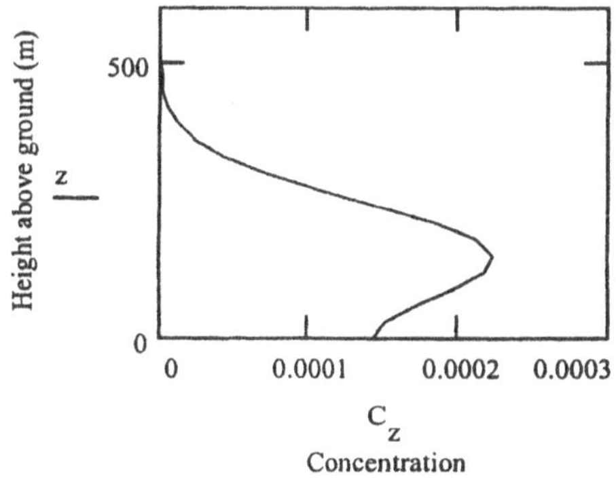

Figure A5 The Influence of Different Atmospheric Stabilities on Ground Level Concentrations

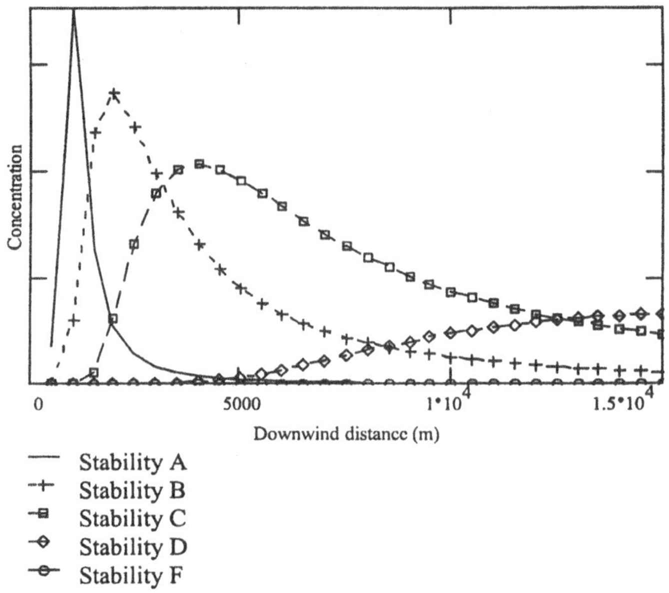

5. The impact of the two sources on ground level concentrations was modelled in a similar manner to the solution to Question 4, and is represented as Figure A6.

Figure A6 Ground Level Impact from an Elevated Source and one at Ground Level

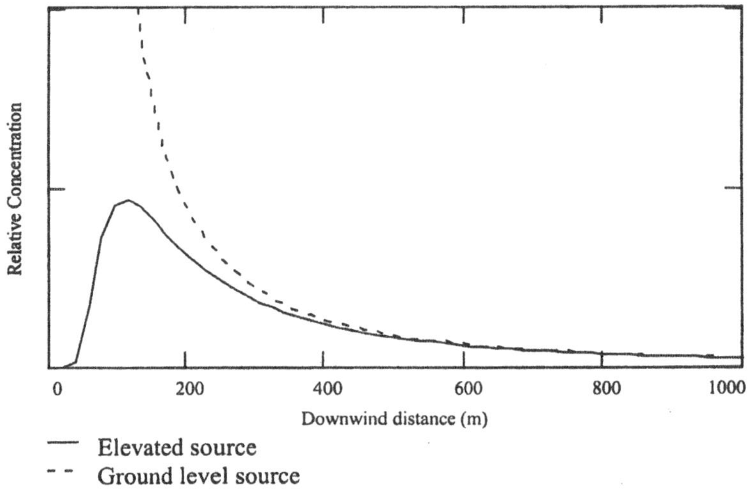

It is evident from Figure A6 that a source at ground level has its greatest impact at ground level close to the source itself, with the concentration declining with distance. This differs from the behaviour of an elevated source that produces the greatest impact some distance from the source.

6. This exercise illustrates several of the uncertainties that can arise in modelling, especially when judgements have to be made. A particular case here is whether the area is urban or rural. The assessment can start by identifying the Pasquill stability category from Table 3.8. Under cloudy conditions with a wind speed of 8 m s^{-1} the category is D.

The calculation of the dispersion coefficients may be made using Figures 5.4 and 5.5 or from the equations provided in Tables 5.1 and 5.3. Bear in mind that the different approaches to estimating these coefficients may give different results. For example, from Table 5.3 for an urban environment:

$\sigma_y = 0.16x(1 + 0.0004x)^{-0.5}$ $\sigma_z = 0.14x(1 + 0.0003x)^{-0.5}$

So: $\sigma_y = 0.16 \times 700(1 + 0.0004 \times 700)^{-0.5}$ $\sigma_z = 0.14 \times 700(1 + 0.0003 \times 700)^{-0.5}$

Hence: $\sigma_y = 99$ m $\sigma_z = 89$

For the rural environment:

$\sigma_y = 0.08x(1 + 0.0001x)^{-0.5}$ $\sigma_z = 0.06x(1 + 0.0015x)^{-0.5}$

From which $\sigma_y = 54$ m and $\sigma_z = 29$ m

From the figures, $\sigma_y = 45$ m and $\sigma_z = 24$ m, which are of similar order to the calculated values for open country. As the site is at the edge of an urban area there is a question over whether rural or urban conditions should be modelled, given the limited information. Consider first an urban environment.

It is necessary to estimate the wind speed at the release height. For stability D in an urban environment suggests an exponent of 0.25. So, substituting in

$$u_s = u_1\left[\frac{h_s}{z}\right]^p \text{ gives } u_{60} = 8\left[\frac{60}{10}\right]^{0.25}$$

From which $u_{60} = 12.5 \text{ m s}^{-1}$

The plume rise and effective stack height can now be calculated. First, the value of F is derived from:

$$F = gvr^2\left(\frac{T_s - T}{T_s}\right) = 9.81 \times 21.5 \times \left(\frac{4.3}{2}\right)^2 \times \left(\frac{418-283}{418}\right) = 315 \text{ m}^4 \text{ s}^{-3}$$

From the text, the plume rise under neutral or unstable conditions in the atmosphere (Pasquill stability categories A to D) may be estimated using the expression:

$$\Delta h = 38.71\left(\frac{F^{0.6}}{u_s}\right)$$

So, substituting in these equations, it is found that $\Delta h = 98$ m. Hence, the effective stack height is therefore 60 + 98 = 158 m.

The remaining unknown in the dispersion equation is the emission rate of sulphur dioxide from use of the distillate fuel. The fuel burning rate is given as 484 tonnes per day and the sulphur content 0.3%. So the emission rate of sulphur is:

$$0.003 \times \frac{484}{24 \times 60 \times 60} = 1.68 \times 10^{-5} \text{ t s}^{-1} = 16.8 \text{ g s}^{-1}$$

The sulphur dioxide emission rate is therefore twice this since the molecular mass of sulphur dioxide is 64 while the atomic mass of sulphur is 32. Hence the required emission rate is 33.6 g s^{-1}.

Now the dispersion equation can be used to predict the ground level concentration at the dwellings. The relevant equation is:

$$C = \frac{Q}{\pi u \sigma_y \sigma_z}\left[\exp\left(-\frac{(H)^2}{2\sigma_z^2}\right)\right]$$

Substituting and using the dispersion coefficients from Figures 5.4 and 5.5 gives:

$$C = \frac{33.6}{\pi \times 12.5 \times 99 \times 89}\left[\exp\left(-\frac{(158)^2}{2 \times 89^2}\right)\right] = 9.7 \times 10^{-5} \times 0.21 = 2 \times 10^{-5} \text{g m}^{-1}$$

This concentration of 20 μg m^{-3} has to be added to the existing air quality levels, and is likely to be significant.

However, calculating for a rural environment gives a wind speed at the release height of 10.5 m s^{-1}, a greater plume rise of 116 m leading to an additional concentration of 2×10^{-15} μg m^{-3}. This additional contribution will clearly be negligible. Indeed, running a simple computer model suggests zero ground level impact under these conditions.

The treatment in this example has ignored some factors. It may be recalled, for example, that the urban heat island shifts the atmospheric stability, so different dispersion coefficients may apply than those given in this answer. Nevertheless it serves to illustrate some uncertainties and their implications on a model outcome.

Chapter 7

1. The data from the AERMOD example, indicate that the second highest 24 hr average concentration of sulphur dioxide in 6.57 μg.m^{-3}. In simple terms, Table 3.5 suggests that the 1 hr average concentration will be five times greater. This suggests a concentration of about 33 μg.m^{-3}.

Chapter 8

1 The weather conditions for this scenario indicate Pasquill category D. the other relevant data are x = 1000 m, y = 100 m, z = 0 m, H = 40 m, u = 4 m s^{-1}.

In addition, t = 5 × 60 and Q = 375 × 2 × 60 = 45 000 g. From figures 5.4 and 5.5, $\sigma_y = \sigma_x = 69$ m and $\sigma_z = 31.5$ m.

Using the equation for an instantaneous point source:

$$C = \frac{2Q}{(2\pi)^{1.5}\,\sigma_x\sigma_y\sigma_z}\exp\left[-\frac{1}{2}\left(\frac{x-ut}{\sigma_x}\right)^2\right]\exp\left[-\frac{1}{2}\left(\frac{H}{\sigma_z}\right)^2\right]\exp\left[-\frac{1}{2}\left(\frac{y}{\sigma_y}\right)^2\right]$$

So:

$$C = \frac{2\times 45000}{15.75\times 69\times 69\times 31.5}\exp\left[-\frac{1}{2}\left(\frac{1000-4\times 5\times 60}{69}\right)^2\right]\exp\left[-\frac{1}{2}\left(\frac{40}{31.5}\right)^2\right]\exp\left[-\frac{1}{2}\left(\frac{100}{69}\right)^2\right]$$

Therefore $C = 8.9 \times 10^{-5}$ g m^{-3}

2. There appears an inconsistency in these requirements, with one requiring modelling a finite duration release (10 min), yet the predicted concentration has to be compared against a 1 hr average concentration. To set a model to provide a 1 hr average yet with a finite release, each receptor will experience a plume for something over 10 min, allowing for plume expansion, and dilution air for the remaining period. On the other hand, setting a model to yield 10 min average values provides a substantial safety factor when compared against a 1 hr health protection standard, which itself is likely to include safety factors. It should also be remembered that the concentration given by the Gaussian equation is a mean over the same time interval as that for which the dispersion coefficients and wind speed represent. Following the convention established by Sutton, a 3 min mean is often assumed, although Turner's *Workbook of Atmospheric Dispersion Estimates* indicates a 10 min averaging time.

3 In solving this problem the key parameters are:

$$x = 1 \text{ km} = 1000 \text{ m}$$
$$y = 0.1 \text{ km} = 100 \text{ m}$$
$$H = 30 \text{ m}$$
$$u = 4 \text{ m.s}^{-1}$$
$$Q = (400 \text{ g s}^{-1})(2 \text{ min})(60 \text{ s.min}^{-1}) = 48000 \text{ g}$$

Overcast conditions day or night indicate Stability Category D from Table 2.9, so using Figures 4.4 and 4.5, σ_y = 69m and σ_z =32m. It is assumed that $\sigma_x = \sigma_y$.

The maximal concentration reaches the residential area when it is transported there by the wind, which occurs after (1000 m/4 m.s^{-1})= 250 s. So, the company management have about 4 minutes to give warning of the incident.

The concentration at the houses can be estimated as:

$$c = \frac{2 \times 48000}{(2\pi)^{1.5} \times 69 \times 69 \times 32} \exp\left[-\frac{1}{2}\left(\frac{1000 - 4 \times 250}{69}\right)^2\right] \exp\left[-\frac{1}{2}\left(\frac{30}{32}\right)^2\right] \exp\left[-\frac{1}{2}\left(\frac{100}{69}\right)^2\right]$$

$$= 0.04 \times 1 \times 0.644 \times 0.35 = 9 \times 10^{-3} \text{ g.m}^{-3} = 9000\ \mu\text{g m}^{-3}.$$

Reference to Table 1.3 shows that this value is very much greater than the short term environmental assessment level for chlorine of 300 μg m^{-3}. Clearly a cause for concern.

4. Possible characteristics may include the following, from which five may have been selected.
   i. Chemicals will often be stored as liquids and so the volume of gas resulting from depressurization will be very large.
   ii. While conventional dispersion calculations are based on releases from fixed point sources such as stacks or defined area and line sources, the sites of accidents are not at fixed locations.
   iii. The change of phase from liquid to gas may be due to various causes, such as a jet entraining air to a pool evaporating.
   iv. The high instantaneous concentration of vapour is likely to give the cloud associated with a release of a relatively high density. This will in turn make the cloud disperse at low level and be subject to topographical effects to a considerable degree.
   v. The time scale is likely to be significant for short period releases, compared with continuous releases typical of most air pollution problems.

5. (i) Estimating the steady-state ground level concentration at the closest edge of the village during the night-time release can be done using the Pasquill stability. Assigning a stability category according to the approach in Chapter 3 suggests that class D is applicable. From the graphs of standard deviations σ_y is estimated as 65m and σ_z is estimated as 32 m. Substituting in the Gaussian equation set up as a Mathcad® file gives:

$$Q := 25 \cdot \frac{gm}{sec} \qquad u := 4 \cdot \frac{m}{sec} \qquad Y := 0 \cdot m \qquad H := 50 \cdot m \qquad Z := 0 \cdot m$$

$$\sigma_y := 65 \cdot m \qquad \sigma_z := 32 \cdot m$$

$$C := \frac{Q}{\pi \cdot \sigma_y \cdot \sigma_z \cdot u} \cdot \exp\left[-\left(\frac{1}{2}\right)\cdot\left(\frac{Y}{\sigma_y}\right)^2\right] \cdot \left[\exp\left[\frac{-H^2}{2\cdot(\sigma_z)^2}\right]\right] \qquad C = 2.822 \cdot 10^{-7} \cdot kg \cdot m^{-3}$$

So the predicted concentration at the edge of the village is 282 μg m^{-3}.
Alternatively, running the model in http://www.industrialhygiene.com/calc/model.html gives values of σ_y = 68 and σ_z = 31.5, with the concentration 263 μg m^{-3}. Both answers are of a similar order, and the differences may reflect the differences in computing the standard deviations.

ii) The concentration at the same time and at a similar distance downwind, but at the location of the school is calculated as follows.

$Q := 25 \cdot \frac{gm}{sec}$ $u := 4 \cdot \frac{m}{sec}$ $Y := 200 m$ $H := 50 \cdot m$ $Z := 0 \cdot m$

$\sigma_y := 65 \cdot m$ $\sigma_z := 32 \cdot m$

$$C := \frac{Q}{\pi \cdot \sigma_y \cdot \sigma_z \cdot u} \cdot \exp\left[-\left(\frac{1}{2}\right) \cdot \left(\frac{Y}{\sigma_y}\right)^2\right] \cdot \left[\exp\left[-\frac{H^2}{2 \cdot (\sigma_z)^2}\right]\right] \qquad C = 2.481 \cdot 10^{-9} \cdot kg \cdot m^{-3}$$

So, the concentration at the school located 200 m from the plume centreline is 100 times lower at 2.5 $\mu g\ m^{-3}$. Again, a concentration of 3.5 $\mu g\ m^{-3}$ was derived by running the model in http://www.industrialhygiene.com/calc/model.html.

iii) The conditions on the following day suggest strong insolation and consequently B class stability. Under these conditions, a plot of the ground level concentration (kg m^{-3}) predicted the following day as a function of the distance from the site boundary to a distance of 10 km downwind is shown in Figure A7. This was obtained using the equations for the dispersion coefficients from Chapter 5.

Figure A7 Variation in Ground Level Concentration with Distance from the Source

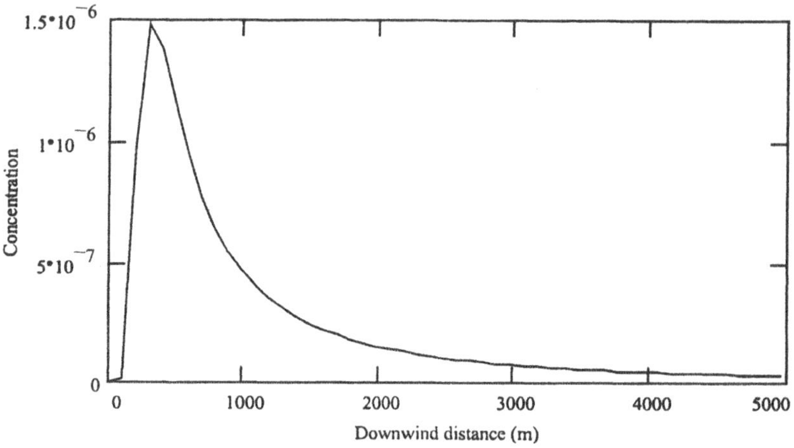

iv) The above plot of concentration (kg m^{-3}) against distance from Mathcad® suggests that the maximum ground level concentration occurs at about 300 m from the source. This is within the site boundary. The concentration is about 1500 $\mu g\ m^{-3}$. This location raises health and safety considerations for those on site rather than health risks to the occupiers of the farm or the village. Running the model in http://www.industrialhygiene.com/calc/model.html gives a maximum concentration at 537 m with a value of 918 $\mu g\ m^{-3}$. This location is close to that predicted from Figure 6.3(a) for a rural setting. While the concentrations are similar within the expectations of modelling, the differences in locations clearly have different implications, with the farm being more at risk.

This example illustrates that mathematical precision must never be confused with accuracy in modelling.

v) Assuming that no ambient air quality standards exist for the released substance, appropriate criteria to help in assessing the significance of the emission include environmental assessment levels, perhaps based on occupational standards. Equally, flammable limits may be appropriate, depending on the nature of the released substance.

6. The release is of finite duration as the input data suggest. In fact the model shows this as follows in Figure A8

Figure A8 Emission Rate from the Release Model

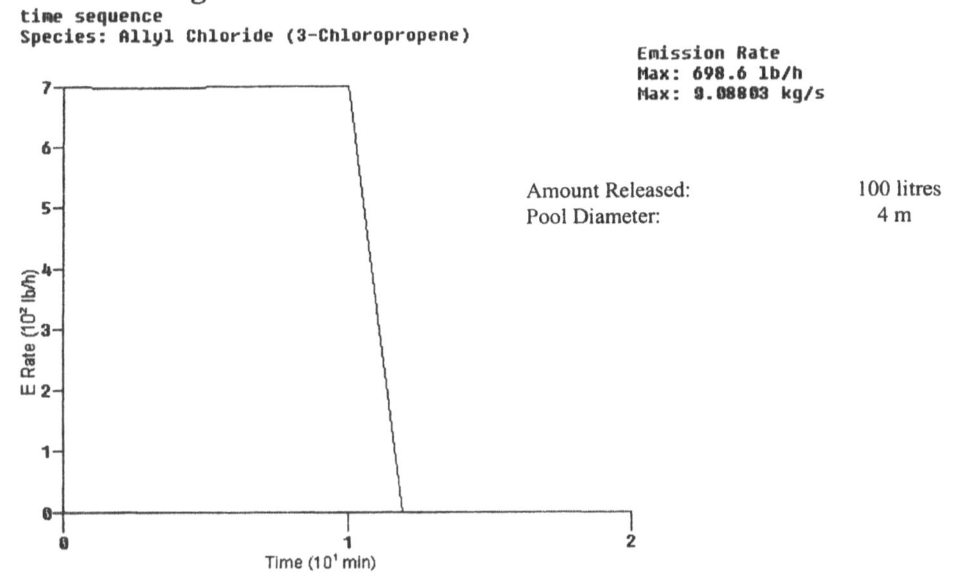

Ignition occurs at ground level with the heat from the burning pool vapour causing the plume to rise. Hence the model output in Figure 8.6 shows no impact at ground level. However, the text on the model output shows that the maximum concentration is 177 m above ground level. A vertical cross section from the model at 1 min after the release reveals plume behaviour shown in Figure A9.

Figure A9 Vertical Plume Cross Section 1 Minute After the Release

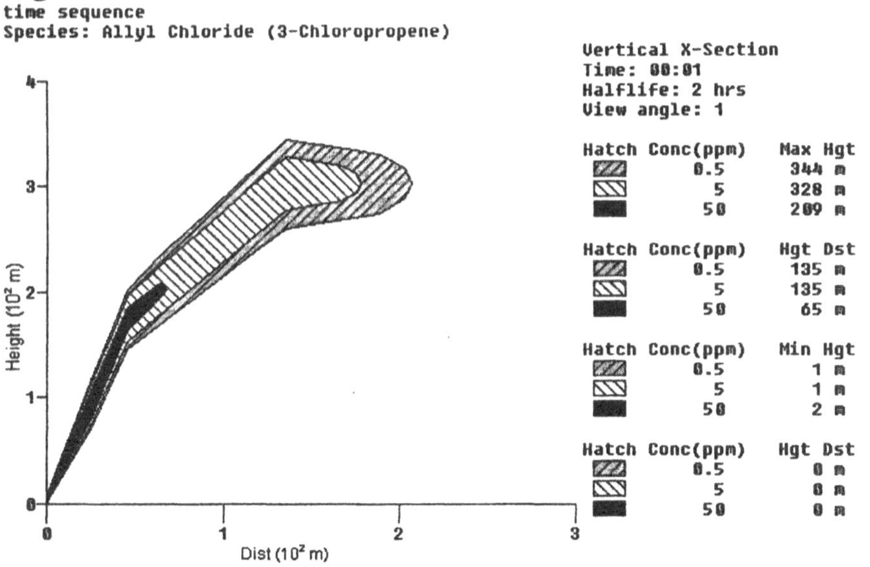

After two minutes, no fuel remains and the plume has completely left contact with the ground and is dispersing through the atmosphere as Figure A10 indicates. After five minutes the plume is dispersing further afield as indicated in Figure A11.

Figure A10 Vertical Plume Cross Section Two Minutes After the Release

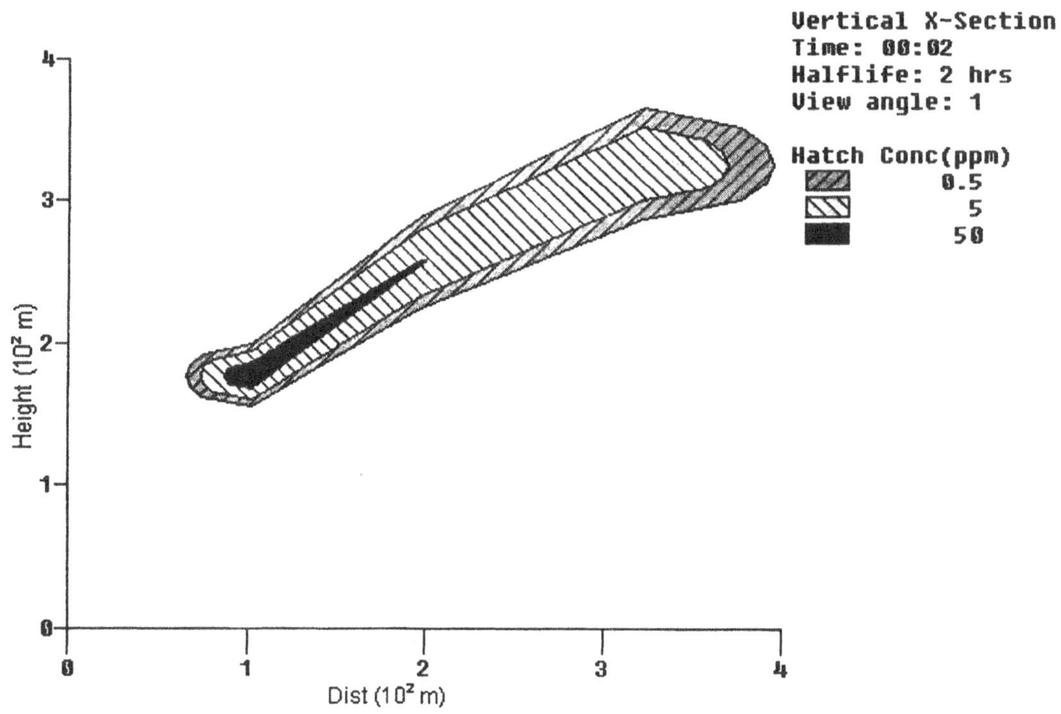

Figure A11 Vertical Plume Cross Section Five Minutes After the Release

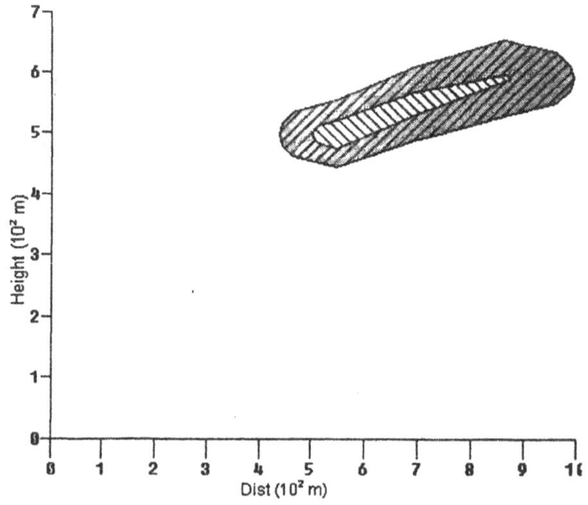

APPENDIX 8
GLOSSARY

Acid deposition	The acids deposited by transfer from the atmosphere by precipitation (rain, fog, snow etc) are wet deposition, while direct transfer to the surface is called dry deposition.
Adiabatic lapse rate	The rate of decrease in temperature with increase in altitude of a bubble of air that expands slowly to the lower atmospheric pressure, without exchange of heat. For a descending bubble, it is the rate of increase in temperature with decrease in altitude. Theory predicts that for dry air the adiabatic lapse rate is 9.8°C km^{-1}. The value is less for moist air and depends on the moisture content of the air.
Advection	The transport of air, its properties (such as heat), trace components, fog, etc. Solely by mass motion of the atmosphere, generally horizontally.
Aerometric measurements	Measurements of the temperature, pressure, air motion (velocity vectors) or other physical characteristics of an air mass.
Aerosol	Dispersions of particles (solid, liquid or mixed) in a carrier gas, usually air.
Air pollution	The presence in the outdoor atmosphere of one or more contaminants, such as dust, fumes, gas, mist, odour, smoke or vapour, in quantities or characteristics, and of duration that may be injurious to human, plant or animal life or to property, or which unreasonably interferes with the comfortable enjoyment of life and property
Air quality	Ambient concentrations of atmospheric components and their temporal and spatial distribution.
Albedo	The fraction of the energy of electromagnetic radiation reflected from a surface relative to the energy incident upon it.
Algorithm	A specific mathematical calculation procedure. A model may contain several algorithms.
Ambient air quality	A general term used to describe the quality of the outside air.
Anticyclone	A large system of winds that rotate about a centre of high atmospheric pressure. When viewed from above, the circulation is clockwise in the northern hemisphere and counter-clockwise in the southern hemisphere.
Area source	An area source is an emission that emanates from a specific, well-defined surface, such as a lagoon or open-top tank.
Atmosphere	The entire mass of air surrounding the Earth, comprising primarily nitrogen, oxygen, water vapour, clouds, carbon dioxide, as well as trace gases and aerosols.
Background	Ambient pollutant concentrations due to natural sources; nearby sources other than the one(s) currently under consideration, and unidentified sources.
Background concentration	Typically the term refers to the air mass in the absence of contributions from a source being modelled or studied.
Boundary Layer	See planetary boundary layer
Calibrate	An objective adjustment using measured air quality data (for example, an adjustment based on least-squares linear regression).
Calm	In air quality modelling, calm is used to define the situation when the wind is indeterminate with regard to speed or direction.

Cavity (Recirculation cavity)	A zone of turbulent airflow above a rooftop created by air flow around the building or nearby buildings and trees. A plume released in a recirculation cavity can easily be pulled back into the same building through open windows or ventilation air intakes, or a plume may reach nearby occupied grounds while the concentrations of contaminants are still excessive.
Chimney	Conventionally, a vertical flue for carrying cooled combustion products of a process to the atmosphere. A legal definition of chimney from the Clean Air Act includes 'structures and openings of any kind from or through which smoke, grit, dust or fumes may be emitted and, in particular, includes flues...'.
Complex terrain	Terrain exceeding the height of the stack being modelled.
Cold front	The transition zone where a cold air mass is replacing a warmer air mass. The air behind a cold front is noticeably colder and drier than the air ahead of it. A cold front is represented on a weather map by a solid line with triangles along the front pointing towards the warmer air and in the direction of movement.
Computer code	A set of statements that make up a computer program.
Concentration	The quantity of matter (mol, g, molecule or other) per unit of volume (cm^3, l, m^3 or other). The mixing ratio is also used as a measure of relative concentration (for example, 1 ppm representing one molecule of a species in a million molecules of air). For a non-reactive gas, the mixing ratio does not change with altitude, but the absolute concentration (molecules. cm^{-3}) decreases.
Diffusion	The spreading or scattering of a gaseous or liquid material. Eddy diffusion in the atmosphere is the process of gas transport due to turbulent mixing in the presence of a composition gradient. Molecular diffusion is the net transport of molecules that results from their molecular motions alone in the absence of turbulent mixing. Eddy diffusion is the most important mixing process in the lower atmosphere, while molecular diffusion becomes significant at lower pressures in the upper atmosphere.
Dose	The amount of a substance available for interaction with metabolic processes or biologically significant receptors after crossing the outer boundary of an organism. The potential dose is the amount ingested, inhaled or applied to the skin. The applied dose is the amount of a substance presented to an absorption barrier and available for absorption (although not necessarily having yet crossed the outer boundary of the organism). The absorbed dose is the amount crossing a specific absorption barrier (for example, the exchange boundaries of skin, lung and the digestive tract) through uptake processes.
Downdraught	The downward motion of a plume as a result of the low pressure in the cavity zone on the leeward side of a building or high ground.
Downwash	As applied to the action of gases from a chimney, it is the downward motion of the gases brought about by eddies that form in the lee of the chimney when a wind blows.
Eddy	In turbulent fluid motion, a bubble of a fluid that has some definitive character and moves in some way differently from the main flow.
Eddy dispersion (or diffusion)	The process by which substances are mixed in the atmosphere or other fluid by eddy motion.
Effective chimney height	A height used for calculating the dispersion of gases emitted from a chimney. It differs from the physical height of the chimney by an amount that depends on factors such as the exit velocity, buoyancy and wind speed. It may also be influenced by the local topography. It denotes the maximum height of the centre of the plume path above the ground level.

Emission	The total rate at which a solid, liquid or gaseous pollutant is released into the atmosphere from a given source. It is usually expressed as mass per unit time.
Emission inventory	A systematic collection of information on air pollutant emissions in a given area. Usually the types of sources and the heights of release are summarized as well as the composition and rate of release. Temporal variations are useful inclusions.
Emission source	There are several types of emission source. Point sources include power plant, incinerators, boilers etc, and often involve release from tall chimneys. Area sources include spills and lagoons, as well as collections of smaller point sources. Mobile sources of emission include traffic, ships, aircraft etc.
Equilibrium	A dynamic or static state of balance between opposing forces or actions.
Evaluate	To appraise the performance and accuracy of a model based on a comparison of concentration estimates with observed air quality data.
Flagpole	Flagpole receptors refer to the height of the receptor being above local ground level.
Flue gas	Waste gas from a combustion process.
Fluid modelling	Modelling done in a wind tunnel or water channel to quantitatively evaluate the influence of buildings and/or terrain on pollutant concentrations.
Flux	A rate of flow through a given unit of surface per unit time.
Frequency distribution	If the range of observed measurement values is sub-divided into a regular sequence of smaller intervals, this distribution is a plot of the frequency of occurrence of values falling into each interval.
Fugitive dust	Dust discharged into the atmosphere in an unconfined flow stream such as that from unpaved roads, storage piles and heavy construction operations.
Fugitive emission	Fugitive emissions are adventitious releases from a source. Sources referred to as having 'fugitive' emissions (for example, multiple leaks within a specific processing area) are typically modelled as area sources.
Fumigation	A phenomenon in which air pollution, retained by an inversion layer near its level of emission, is brought rapidly to ground level as the inversion breaks up. Fumigation occurs when a plume is emitted into a stable layer of air and that layer is subsequently mixed to the ground either through convective transfer of heat from the surface or because of advection to less stable surroundings. Fumigation may cause excessively high concentrations but is usually rather short-lived at a given receptor. Fumigation is also an important phenomenon on and near the shoreline of bodies of water. This can affect both individual plumes and area-wide emissions.
Gaussian function	A distribution function defined by the Gaussian equation.
Geostrophic wind	The wind blowing above the boundary layer that, by its strength and direction, represents the balance between the pressure–gradient force, acting directly from the region of higher pressure towards the region of lower pressure, and the Coriolis force (effect), deflecting moving air to the right (in the northern hemisphere and to the left in the southern hemisphere). When these are in balance the wind flows parallel to the isobars.
Gradient wind	Wind that represents the balance of all the forces acting upon moving air.
Ground level inversion	An inversion of the usual temperature gradient in the atmosphere; the temperature of the air increases with altitude. This leads to poor mixing of gases released beneath the inversion.
Gustiness	Intensity of turbulence.
Inversion	A departure from the normal decrease of temperature with increasing altitude in the lower atmosphere. Several inversions may be present at different altitudes in a given air mass as a result of meteorological factors.

Inversion height	The height above ground level at which there is a change in the sign of the normal lapse rate.
Isobar	Lines on a weather map joining points of equal barometric pressure.
Isopleth	A line connecting points of equal concentration around a plume.
Lapse rate	The variation of an atmospheric variable with height usually related to temperature. The lapse rate is taken as positive when the temperature decreases with height.
Mixing height	The height to which significant mixing occurs of pollutants releases into the atmosphere.
Mixing ratio	In meteorology, the dimensionless ratio of the mass of a substance in an air bubble to the mass of the remaining substances in the bubble. In atmospheric chemistry, the term is used to represent concentrations of trace substances by volume as well as by mass.
Model	A quantitative or mathematical representation or simulation which attempts to describe the characteristics or relationships of physical events.
Monin–Obukhov length	The distance above ground level at which mechanically derived turbulence equals the buoyancy-derived turbulence.
Monte Carlo technique	A repeated random sampling from the distribution of values for each of the parameters in a generic equation (for example, for exposure or dose) to derive an estimate of the distribution in the population.
Occluded front	A developing cyclone typically has a preceding warm front (the leading edge of a warm moist air mass) and a faster moving cold front. As the storm intensifies, the cold front rotates around the storm and catches the warm front. This forms an occluded front, when a cold front overtakes a warm front. A noticeable wind shift also occurs across an occluded front.
Planetary boundary layer	The lower part of the atmosphere between the surface and the level of the geostrophic wind. It is the well-mixed region of the lower atmosphere in which turbulence is maintained largely by convective buoyancy. This is induced by the upward heat flux originating from the solar-heated surface of the Earth. During the afternoon, this may extend from 1–5 km in height.
Plume	A mass of air and contaminant released from a source spreading out as it moves away from the source. The shape of the plume may assume various geometrical forms and concentration distributions according to the topography and meteorology.
Point source	A point source is an emission that emanates from a specific point, such as a chimney or vent.
Potential temperature	The temperature that a dry air bubble would have if lowered or raised adiabatically to a level of 1000 mbar pressure.
Preferred model	A model that is recommended for a specific type of regulatory application according to US legislation. See also Regulatory model.
Receptor	A location at which ambient air quality is measured or estimated.
Receptor models	Procedures that examine an ambient monitor sample of particulate matter and the conditions of its collection to infer the types or relative mix of sources impacting on it during collection.
Receptor	A location at which ambient air quality is measured or estimated.
Refined model	An analytical technique that provides a detailed treatment of physical and chemical atmospheric processes and requires detailed and precise input data. The estimates are more accurate than those obtained from conservative screening techniques.
Regulatory models	A term normally used for models that can be used to determine whether a polluter complies with certain rules and regulations for air quality management. An alternative term is compliance or preferred models.

Release height — The height above the ground at which contaminants are released. This may be the effective chimney height but will often be the roof height of a building plus the effective height of a chimney above the roof of that building.

Rollback — A simple model that assumes that if emissions from each source affecting a given receptor are decreased by the same percentage, ambient air quality concentrations decrease proportionately.

Screening technique — A relatively simple analysis technique to determine if a given source is likely to pose a threat to air quality. Concentration estimates from screening techniques are conservative.

Simple terrain — An area where terrain features are all lower in elevation than the top of the stack of the source.

Stability — In meteorology, the stability of the atmosphere in the vertical direction to vertical displacements. If Γ is the dry adiabatic lapse rate, dry air is stable if the environmental lapse rate $< \Gamma$, unstable if $> \Gamma$ and neutral if it $= \Gamma$.

Stack — See chimney.

Stagnation — Stagnation conditions are characterized by calm or very low wind speeds and variable wind directions. These conditions may persist for several hours to several days. During stagnation conditions there is poor dispersion of air pollutants, especially from low-level emissions sources. This can lead to relatively high ground-level concentrations. Stagnant conditions, including extended periods of calm, often produce high concentrations over wide areas for relatively long periods. Standard short term Gaussian models are often not applicable to such situations.

Troposphere — The lowest layer of the atmosphere, ranging from ground level to the base of the stratosphere at 10–15 km altitude, depending on latitude and meteorology.

Turbulence — Atmospheric eddies, mechanical or thermal, which are produced primarily by shearing stress or by convection.

Warm front — The transition zone where a warm air mass is replacing a cold air mass. The air behind a warm front is warmer and moister than the air ahead of it. The front is represented on a weather map by a solid line with semicircles pointing towards colder air and in the direction of movement. The lighter warm moist air behind the front can be lifted upward and 'override' the colder air. As the air rises, it cools, and if enough water vapour condenses, widespread clouds and precipitation develop. The frontal surface gently slopes up and over the cold air mass, so the upward motions associated with warm fronts are typically weaker than those ahead of a cold front.

Wind rose — A diagram designed to show the distribution of wind direction at a location over a considerable period of time. Usually shown in polar coordinates, with the distance from the origin being proportional to the probability blowing from the given angle, usually measured from the north.

Worst case — A semi-quantitative term referring to the maximum possible exposure, dose or risk that can conceivably occur, whether or not this exposure, dose or risk actually occurs in a specific population.

INDEX